Practical Impedance Matching Techniques, Software and scripts.

Ain Rehman
Signal Processing Group Inc

Copyright 2012, 2016, 2017, 2018 by Signal Processing Group Inc.

This eCADbook© is published by Signal Processing Group Inc.

By Ain Rehman.

All rights reserved including the right of reproduction in whole or in part in any form.

Manufactured in the United States of America.

First edition. 2018.

Practical Impedance Matching Techniques

In the design of radio frequency or wireless circuits the concept of power and power transfer is key. Radio and high frequency parameters are usually based on power quantities unlike that in the low frequency regime where voltages and currents are usually deemed appropriate and sufficient. Power transfer from the source to the load is so important that a number of parameters and critical concepts have evolved to analyze the consequences of this important factor. Among these are the Standing Wave Ratio (Voltage Standing Wave Ratio, Current Standing Wave Ratio and Power Standing Wave Ratio). Reflection Coefficients, Return Loss, Mismatch loss etc. All are related to each other and are in frequent use in the high frequency design community. The understanding of these quantities and their interplay is an essential part of both the practicing engineer and the student's repertoire. Without a solid grounding in these principles the practitioner's art becomes difficult to say the least. This eCADbook™ is another effort to present these important concepts. The difference is, that this book chooses the path of a concise approach where extended mathematical derivations are downplayed and the concepts and parameters are presented in a cookbook fashion. However, mathematics is not completely ignored, but is embodied in a number of CAD routines and scripts (Javascript, C++/Windows) which the reader is encouraged to use in the reading of the book and in subsequent work. In this way the "wood is not obscured by the trees" and the reader can try examples from the book or those in his/her own work to allow a deeper understanding of the principles. Also time is saved by the use of these software tools.

Part one deals with brief intuitive descriptions of the quantities in question. This work lays the foundation for part two which enunciates the mathematical expressions and formulas for these parameters. Part three addresses the lumped element impedance matching approach to power transfer. The concept of impedance matching is very important since high frequency design sometimes becomes the art of matching impedances for proper operation of the circuit blocks and antennas. Part four continues the impedance matching discussion using the concept of the Smith Chart and transmission lines. In this case the matching elements can be deemed distributed. There is significant discussion of discrete lumped elements and transmission line quantities including microstrip lines.

For each topic under discussion, there are a number of CAD routines and scripts included. Finally it should be understood that *not every impedance matching technique* is addressed in this book. Only the more popular ones. However the assumption is that once one becomes familiar with the techniques presented here further reading and research may become easier to understand, both intuitively and in terms of mathematical approaches.

CONTENTS

Part I <u>The basics.</u>

1.1   Propagation on transmission
      lines   11
      Standing waves, the reflection coefficient
      and the Standing Wave
      Ratio   12

Part II   <u>Formulas and mathematical interpretations.</u>

      VSWR   18
      Reflection coefficient   19
      Return loss   20
      Input return loss   22
      Output return loss   23
      Voltage reflection coefficient   23
      Voltage standing wave ratio   24
      Relationship of the VSWR to
      the return loss   24
      Mismatch loss   28
      Maximum power transfer
      rule   30

Part III   <u>Lumped element impedance matching techniques</u>
              34

3.1   Definition of the impedance matching problem
              34
3.2   Narrow band pi, L and T impedance matching
              networks   35

3.3	The Q matching technique	33
3.4	Impedance matching of complex terminations	39
3.5	Bandwidth considerations	46
3.6	Using RF transformers and baluns for matching	50
I.1	Transformer symbols	51
I.2	Transformer dot convention	52
I.3	Basic transformer equations	53
I.4	Basic transformer circuit model	54
I.5	Transformer insertion loss	55
I.6	Amplitude and phase imbalance	57
I.7	Configurations of RF transformers Baluns	57
	A simple balun	62
	Transmission line baluns	64
	Guanella balun	65
	Ruthroff balun	68
	Marchand balun	86
	Even and Odd mode impedance	71

Part IV	Transmission line impedance matching techniques	
4.1	Cascaded line matching techniques	73
	Definition of electrical length	74
	Definition of the wave number	75
4.2	The quarter wave transformer	76
	Frequency response of a quarter wave transformer	78
4.3	Multisection transformer impedance matching	79
4.4.2	Percentage bandwidth for the binomial transformer	83
4.4.3	Multisection matching using a Chebyshev transformer	84
4.5	Transmission line expressions and formulas	89
4.5.1	Transmission line facts	89
4.5.2	Electrical length	91
Part V	The Smith Chart and impedance matching techniques	112
Part VI	Appendices and Index	171

Part I

<u>The Basics</u>

A signal propagates on wires or transmission lines that connect a source of power to the load which uses this power to do useful work. In so doing a number of electronic phenomena occur, such as reflection of the input signal that causes a loss of useful power which sometimes dissipates harmlessly or in some critical cases disastrously by generating reflective effects that can destroy the circuit or the equipment it is included in.
To understand and quantify these concepts a number of measures have been developed over time that are in common use throughout the design, standards and manufacturing community, so that a common framework exists, and is accepted by, and understood by, the majority of practitioners in the industry. These basic concepts are described below.

1.1   <u>Propagation on transmission lines</u>

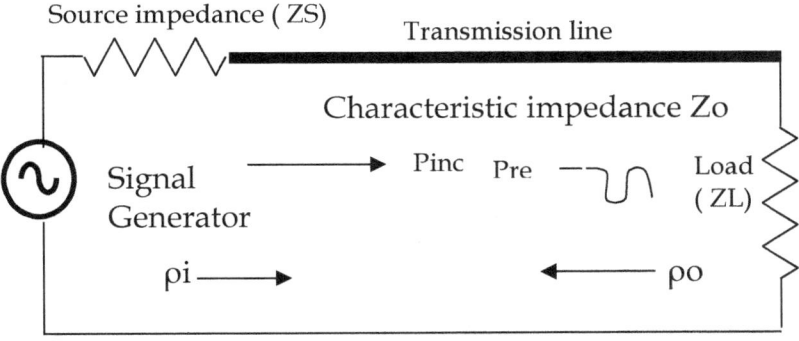

Figure B.1.0

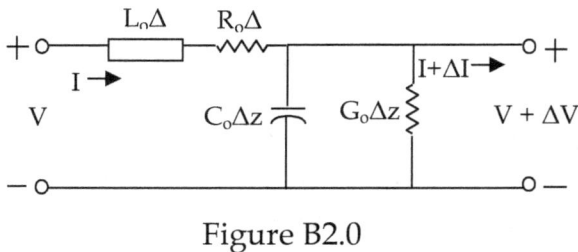

Figure B2.0

Lo, Ro, Co and Go are constants of the line and are evaluated per unit length. The total values of these elements are then the product of the length of the line and the elementary constants.

V and I are the voltage and current traveling on the line. The incremental voltage and current is also shown in Figure B2.0. These may be deemed to be intermediate values and again can be evaluated over the total length of the line.

1.2 Standing waves, the reflection coefficient and the Standing Wave Ratio.

In addition to this, if the terminations on the line (ZS, ZL) are not *exactly conjugately matched,* reflection of the incident signal (Pinc) occurs at the load end, and a reflected signal (Pref) travels back towards the source. This can set up standing waves on the line as shown below. ρi and ρo are the reflection coefficient magnitudes. The reflection coefficient is defined as the *ratio of the reflected power by the incident power.* ρi is the reflection coefficient magnitude for the reflection of source power with a mismatched transmission line characteristic impedance and ρo is the load end reflection coefficient that represents the mismatch with the load and the characteristic impedance of the line.

The characteristic impedance Zo is the uniform impedance of the line that depends on the line's physical constants and the dielectric medium in which the line is resident. In case of microstrip lines the characteristic impedance depends on the width, height and effective dielectric constant of the microstrip line. More detail on this is presented later on.

An understanding of the reflection coefficient and the VSWR can be facilitated by using the following equation (presented without proof) and using a simple program to generate a graphical look at standing waves.

$$V(t, x) = A\sqrt{4\rho\cos^2 \beta x + (1-\rho)^2}\ \cos(\omega t+\theta), \quad (B1.0)$$

where ρ is the reflection coefficient magnitude, β is the wave number (angular frequency/wave phase velocity; alternatively $2\pi/\lambda$, where λ is the wavelength) and θ is the phase and is given by $\dfrac{(1+\rho)}{(1-\rho)}\cot(\beta x)$. As can be seen the voltage is both a function of x, the distance and the time.

The voltage variation at any point on a transmission line with arbitrary loads is:

$$V = A\sqrt{4\rho\cos^2 \beta x + (1-\rho)^2} \quad (B2.0)$$

where A is the signal amplitude. This equation can be implemented in a simple computer script and the results can be plotted to give a glimpse into the behavior of standing waves. A true plot would include both the time variation, as well as variation with distance. However, a plot of the standing waves with respect to distance is good enough to further the understanding of standing waves and the VSWR.

In the plots shown we have set $\beta = 1$ and the amplitude $A = 1$ without loss of generality since our intention is to look at the effect of ρ on the standing waves. Figure B3.0 shows the case for a reflection coefficient of 1.0.

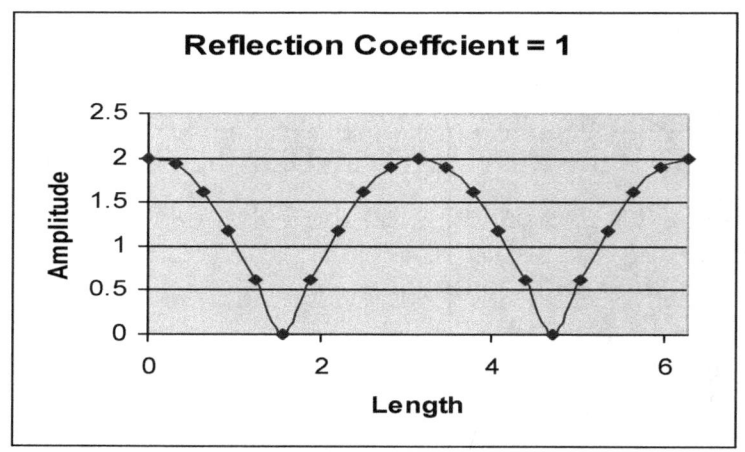

$\rho = 1$. Complete reflection, SWR = infinite : Figure B3.0

Figure B4.0 shows the case when the incident signal is completely dissipated by the load. In this case there is no reflected signal and the VSWR is 1.0, i.e, a perfectly matched case.

Practical Impedance Matching Techniques

Figure B5.0 shows the case of $\rho = 0.5$, while figure B6.0 shows the case for $\rho = 0.2$. From these figures and analysis, a graphical understanding of the effect of the reflection coefficient can be developed. The reader may run his own plots with different values of the reflection coefficient using the GNUPLOT scripts as shown. Just type in the script using the plot command <script> and make sure that the sampling rate is at least 1000. Using this technique, the effects of reflection can be analyzed in detail.

The SWR is simply the maximum value of the amplitude divided by the minimum value of the amplitude.

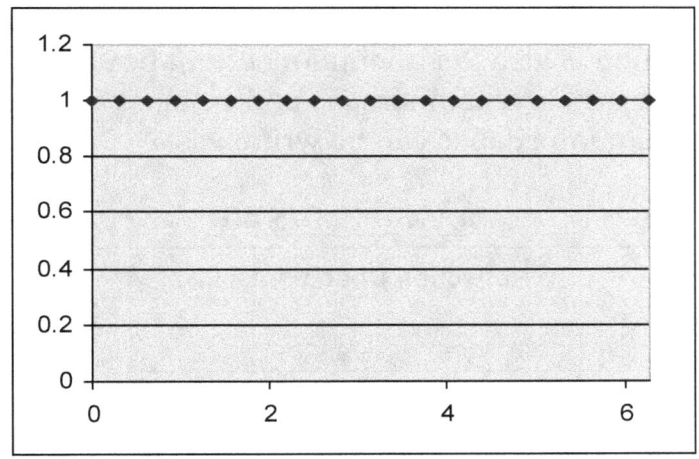

$\rho = 0$. Complete absorption, SWR = 1: Figure B4.0

The plots are generated with the length of line in radians along the x – axis, and the amplitude along the y – axis. In these plots the incident voltage amplitude is set to unity and the wave number is also set to unity for clarity.

Generally the measurements of this type are done using the slotted line technique. The slotted line is simply a transmission line containing a movable probe. The movable probe can be set at various positions along the line. The probe measures the voltages at these positions. Knowing the voltage maxima and minima with length, a number of important parameters can be found, such as the reflection coefficient, VSWR, value of the unknown load impedance and so on.

The characteristics of the current along the line can also be found. It should be obvious that, where on the length of line, the traveling waves add to generate the voltage maxima, they also subtract to generate the current minima. The maximum voltage position is also the minimum current position. The impedance at the current minima is at the maximum impedance. This maximum impedance can be written as:

$$ZMAX = Zo.VSWR.$$

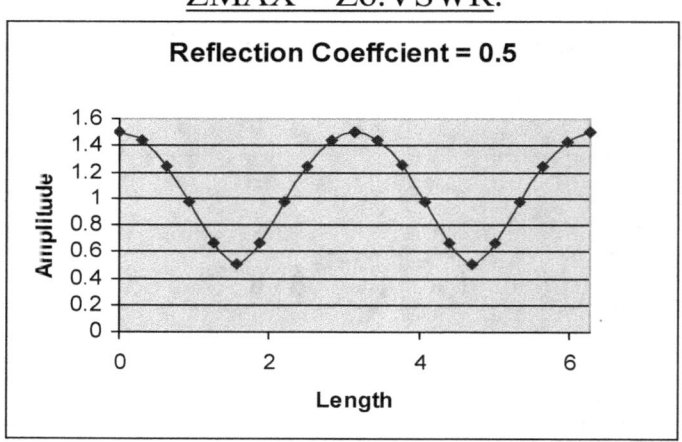

$\rho = 0.5$. Partial reflection, SWR = 3.0 :
Figure B5.0

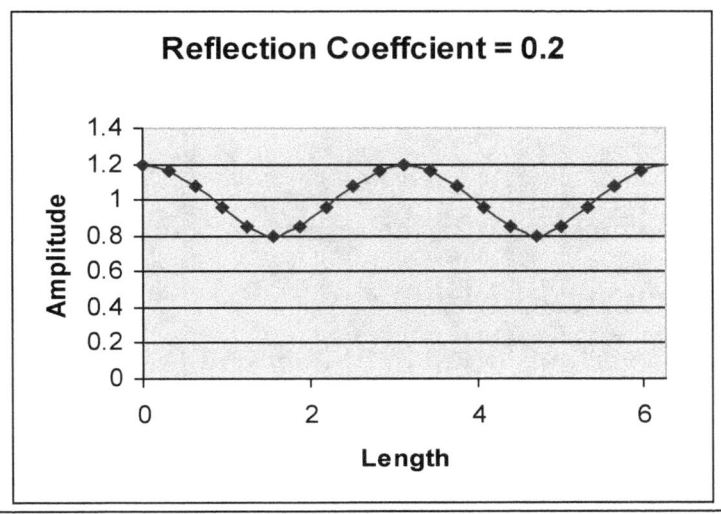

ρ = 0.2. Partial reflection, SWR = 1.5 : Figure B6.0

With this level of understanding we can move on to the next part of the eCADbook™ and look at various other quantities and parameters involved

Part II
<u>Parameters and expressions</u>

<u>VSWR, Reflection coefficient, Return loss, s11/s22.</u>

2.1   VSWR:

The SWR is usually defined as a voltage ratio called the **VSWR**, for *voltage standing wave ratio*. For example, the VSWR value 1.2:1 denotes a maximum standing wave amplitude that is 1.2 times greater than the minimum standing wave value. It is also possible to define the SWR in terms of current, resulting in the ISWR, which has the same numerical value. The *power standing wave ratio* (PSWR) is defined as the square of the VSWR.

The VSWR is related to the reflection coefficient as:

$$\text{VSWR} = \frac{V_{max}}{V_{min}} = \frac{1+\rho}{1-\rho}$$

where ρ = the magnitude of the reflection coefficient.

2.2 Reflection coefficient:

Reflections occur as a result of discontinuities, such as an imperfection in an otherwise uniform transmission line, or when a transmission line is terminated with other than its characteristic impedance. The reflection coefficient Γ is defined thus:

$$\Gamma = \frac{V_r}{V_f}$$

Γ is a complex number that describes both the magnitude and the phase shift of the reflection. The simplest cases, when the imaginary part of Γ is zero, are:

- $\Gamma = -1$: maximum negative reflection, when the line is short-circuited.
- $\Gamma = 0$: no reflection, when the line is perfectly matched.
- $\Gamma = +1$: maximum positive reflection, when the line is open-circuited.

For the calculation of VSWR, only the magnitude of Γ, denoted by ρ, is of interest. Therefore, we define:

$$\rho = |\Gamma|.$$

2.3 Return loss:

Return loss or **Reflection loss** is the reflection of signal power resulting from the insertion of a device in a transmission line or optical fiber. It is usually expressed as a <u>ratio in dB relative to the transmitted signal power</u>.

If the power transmitted by the source is P_T and the power reflected is P_R, then the return loss in dB is given by:

$$RL(dB) = 10\log_{10}\frac{P_T}{P_R}$$

Return Loss is a positive number, historically ORL has also been referred to as a negative number. Within the industry expect to see RL referred to variably as a positive or negative number.

This RL sign ambiguity can lead to confusion when referring to a circuit as having high or low return loss; so remember:- High Return Loss = lower reflected power = large RL number = generally good. Low Return Loss = higher reflected power = small RL number = generally bad.

In metallic conductor systems, reflections of a signal traveling down a conductor can occur at a discontinuity or impedance mismatch. <u>The ratio of the amplitude of the reflected wave V_r to the amplitude of the incident wave V_i is known as the reflection coefficient Γ</u>.

$$\Gamma = \frac{V_r}{V_i}$$

When the source and load impedances are known values, the reflection coefficient is given by:

Practical Impedance Matching Techniques

$$\Gamma = \frac{Z_L - Z_S}{Z_L + Z_S}$$

where Z_S is the impedance toward the <u>source</u> and Z_L is the impedance toward the <u>load.</u>

Return loss is simply <u>the magnitude of the reflection coefficient in dB</u>. Since power is proportional to the square of the voltage, then <u>return loss</u> is given by:

$$RL(dB) = -20Log(|\Gamma|)$$

where the vertical bars indicate magnitude. Thus, <u>a large positive return loss indicates the reflected power is small relative to the incident power, which indicates good impedance match from source to load.</u>

When the actual transmitted (incident) power and the reflected power are known (i.e. through measurements and/or calculations), then the return loss in dB can be calculated as the difference between the incident power P_i (in dBm) and the reflected power P_r (in dBm).

$$RL(dB) = P_i(dBm) - P_r(dBm)$$

s11/s22 relationship to impedance matching:

2.4 Input return loss

Input return loss (RL_{in}) is a scalar measure of how close the actual input impedance of the network is to the nominal system impedance value and, expressed in logarithmic magnitude, is given by:

$$RL = |20Log_{10}(|s11|)| \text{ dB}.$$

By definition, return loss is a positive scalar quantity implying the 2 pairs of magnitude (|) symbols. The linear part, |s11| is equivalent to the reflected voltage magnitude divided by the incident voltage magnitude.

2.5 Output return loss

The output return loss (RL_{out}) has a similar definition to the input return loss but applies to the output port (port 2) instead of the input port. It is given by:

$$RL_{out} = |20 Log_{10}(|s22|)| dB.$$

2.6 Voltage reflection coefficient

The voltage reflection coefficient at the input port (ρ_{in}) or, at the output port (ρ_{out}) are equivalent to s11 and s22 respectively, so:

$$\rho_{in} = s11 \text{ and } \rho_{out} = s22$$

As s11 and s22 are complex quantities, so are ρ_{in} and ρ_{out}.

Voltage reflection coefficients are complex quantities and may be graphically represented by polar diagrams on Smith Charts.

2.7 Voltage standing wave ratio

The voltage standing wave ratio (VSWR) at a port, represented by the lower case 's', is a similar measure of port match to return loss but is a scalar linear quantity, the ratio of the standing wave maximum voltage to the standing wave minimum voltage. It therefore relates to the magnitude of the voltage reflection coefficient and hence to the magnitude of either s11 for the input port or s22 for the output port.

At the input port, the VSWR (Sin) is given by:

$$\text{Sin} = \frac{1+|s11|}{1-|s11|}$$

At the output port, the VSWR (Sout) is given by:

$$\text{Sout} = \frac{1+|s22|}{1-|s22|}$$

2.8 Relationship of VSWR to Return Loss:

VSWR and return loss are related quantities. Note that the reflection coefficient Γ can be written in terms of the VSWR as:

$$\Gamma = \frac{VSWR-1}{VSWR+1}$$

This expression can be derived as follows:

From the given relationship,

$$\text{VSWR} = \frac{V_{max}}{V_{min}} = \frac{1+\rho}{1-\rho}$$

We can show that ρ is given by:

$$\rho = |\Gamma|,$$

thus,

$$\text{VSWR} = \frac{1+\Gamma}{1-\Gamma}$$

Which provides the relationship we started with.

The return loss can then be cast in terms of VSWR as:

$$RL(DB) = -20\log_{10}[VSWR-1]/[VSWR+1]$$

Conversely,

$$VSWR = \frac{\left[10^{\frac{RL(dB)}{20.0}}\right] + 1.0}{\left[10^{\frac{RL(dB)}{20.0}}\right] - 1.0}$$

So if either of the two quantities is known the other can be calculated from it.

Table 1.0 shows these conversions below for convenience.

This table was generated by using the conversion equations shown above.

Table 1.0

Return loss (db)	VSWR	Reflection coefficient
0	Infinite	1.0
1	17.39	0.891
2	8.724	0.794
3	5.848	0.707
4	4.419	0.630
5	3.569	0.562
6	3.009	0.501
7	2.614	0.446
8	2.322	0.398
9	2.099	0.354
10	1.924	0.316

11	1.784	0.281
12	1.670	0.251
13	1.576	0.223
14	1.498	0.199
15	1.432	0.177
16	1.376	0.158
17	1.328	0.141
18	1.288	0.125
19	1.252	0.112
20	1.222	0.100
20.8	1.195	0.089
21.7	1.179	0.082
22.6	1.16	0.074
23.1	1.15	0.069
23.7	1.139	0.065
24.3	1.129	0.060
24.9	1.120	0.056
25.7	1.109	0.051
26.4	1.100	0.047
27.3	1.109	0.043
28.3	1.079	0.038
29.4	1.07	0.033
30.7	1.06	0.029
32.3	1.049	0.024
34.1	1.04	0.019
36.6	1.03	0.014
40.1	1.019	0.009
46.1	1.009	0.004

It is instructive to examine a graphical view of the relationship between VSWR and the return loss presented below.

Note the slow variation of the return loss as the VSWR reaches between 1.0 to 2.0. Conversely for VSWR of 7 or 8 the return loss is low.

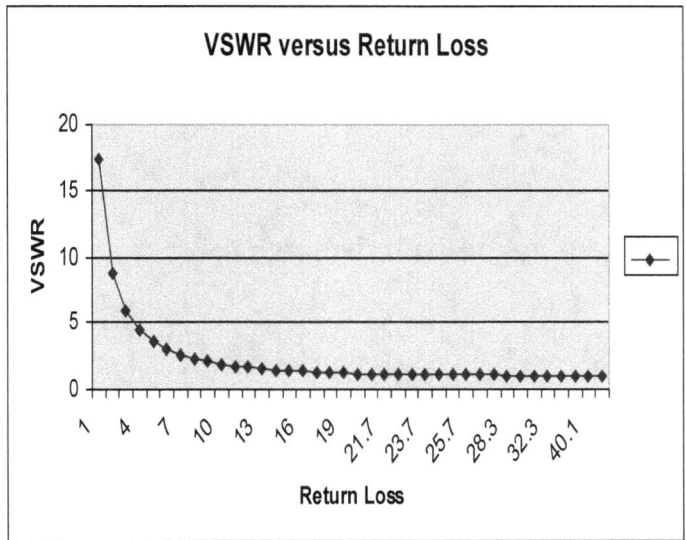

2.9   Mismatch loss:

Mismatch loss in transmission line theory is the amount of power expressed in decibels that will **not be available** at the output due to impedance mismatches and reflections.

Mismatch loss (ML) is the ratio of incident power to the difference between incident and reflected power:

$$ML_{dB} = 10 Log_{10}\left(\frac{P_i - P_r}{P_i}\right)$$

$P_r = P_i - P_d$

where

P_i = incident power
P_r = reflected power
P_d = delivered power

The amount of incident power not reaching the load due to mismatching is:

$$\frac{P_d}{P_i} = 1 - \rho^2$$

where ρ is the reflection coefficient

If the reflection coefficient is known, mismatch can be calculated by:

$$ML_{dB} = 10 \log_{10}(1 - \rho^2)$$

In terms of the voltage standing wave ratio (VSWR):

$$ML_{dB} = 10 \log_{10}\left(1.0 - \left(\frac{VSWR-1}{VSWR+1}\right)^2\right)$$

It is obvious from these expressions that mismatch loss can be calculated from VSWR and the reflection coefficient and vice versa.

2.1.1   Maximum power transfer rule:

Impedance matching depends, to a large extent, on the concept and rule of maximum power transfer.

Practical Impedance Matching Techniques

Simply stated, this theorem states, that a maximum amount of power will be transferred in a electric circuit from the source to the load when the source resistance is equal to the load resistance (in the case of resistive loads) or, that if the impedances are complex, the load impedance is the complex conjugate of the source.

To further illustrate this concept assume a simple circuit like that shown below in Figure PT 1.0

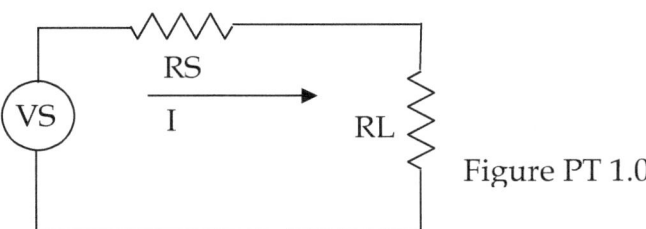

Figure PT 1.0

In this circuit, VS is the source, RS is the source resistance and RL is the load resistance.

The current flowing through the circuit is I. This is a simple circuit but will serve to illustrate the principle of power transfer.

$$I = \frac{VS}{RS + RL} \qquad \text{PT 1.0}$$

The power in the load resistor is $I^2 R$. Then,

Power = $\dfrac{VS^2 RL}{RS^2 + RL^2 + 2RSRL}$ 　　　PT 2.0

This leads to:

Power = $\dfrac{VS^2}{RS^2/RL + RL + 2RS}$ 　　　PT 3.0

If now we let RL be a multiple of RS,

RL = αRS

Then,

Power = $\dfrac{VS^2}{RS/\alpha + \alpha RS + 2RS}$ 　　　PT 4.0

which leads to

Power = $\dfrac{\alpha VS^2}{RS(1+\alpha)^2}$ 　　　PT 5.0

Using this equation we can plot a characteristic of power delivered to the load from the source for various α. This graph is shown below in Figure PT 2.0

Practical Impedance Matching Techniques

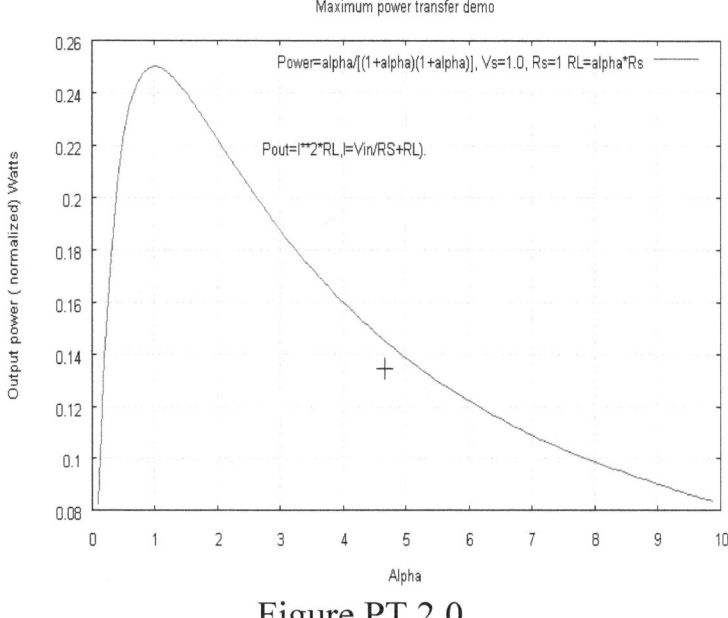

Figure PT 2.0

Note the maximum power at the output occurs at alpha = 1.0. i.e RL = RS.
Also note that the power output in watts is 0.25 Watts or -6.0 dBm.

Part III
Lumped element impedance matching techniques

3.1   Consider Figure 1.0 below. It represents the impedance matching problem.

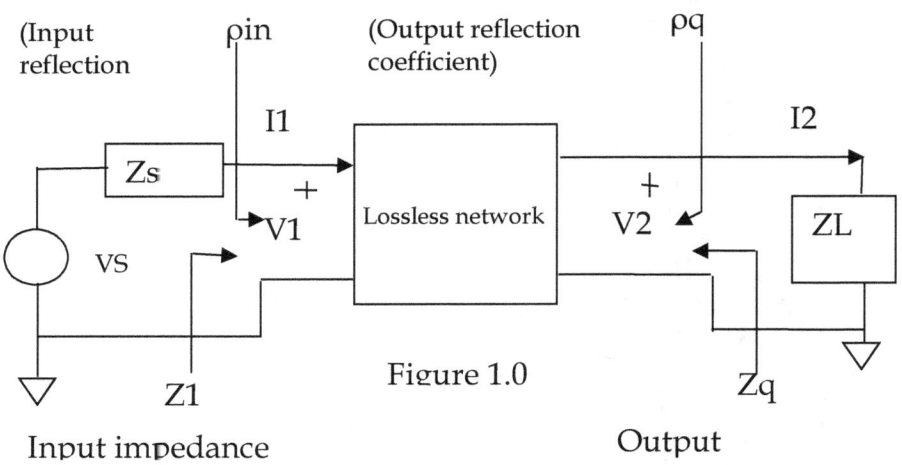

Figure 1.0

The art of impedance matching lies in the design of a network so that a terminating impedance is transformed exactly to a desired impedance at a frequency, or is transformed approximately over a band of frequencies.

An important concept is that:

(a) $Z1 = Zs^*$ (1)

and

(b) $Z2 = Zq^*$ (2)

Also the load impedance must be the complex conjugate of the source.

3.2 Narrow band pi, L and T networks.
The simplest matching networks are the pi, L and T networks shown below. The circuits are composed of reactances as shown. R1 is the input resistance and R2 the output resistance.

Practical Impedance Matching Techniques

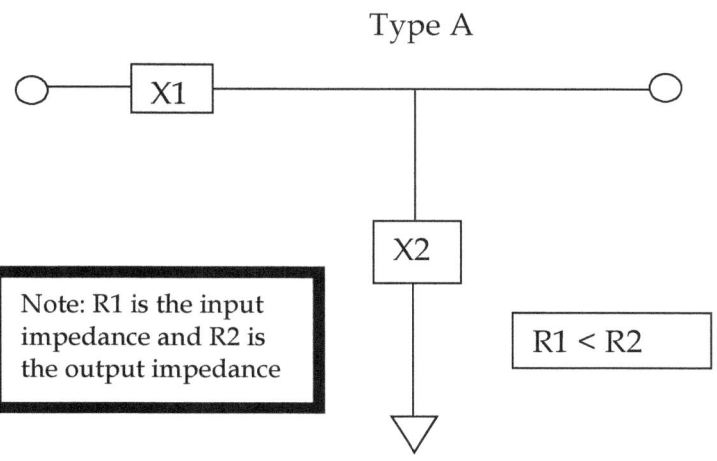

Type A

Note: R1 is the input impedance and R2 is the output impedance

R1 < R2

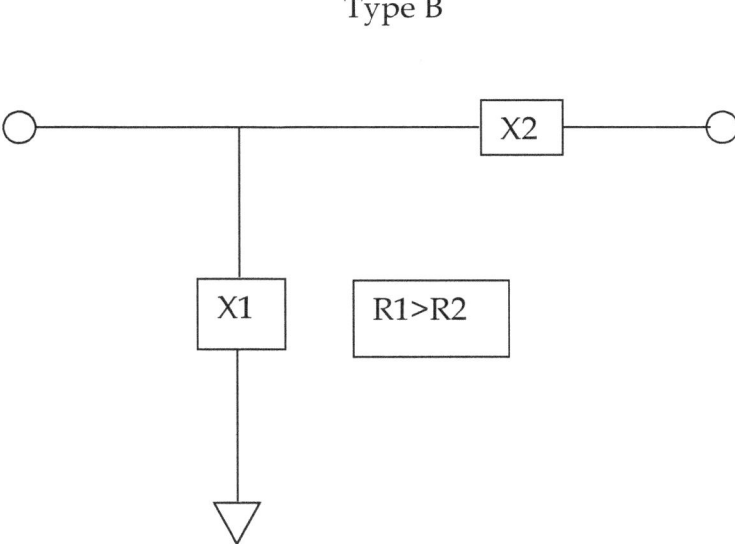

Type B

R1 > R2

Practical Impedance Matching Techniques

Type T

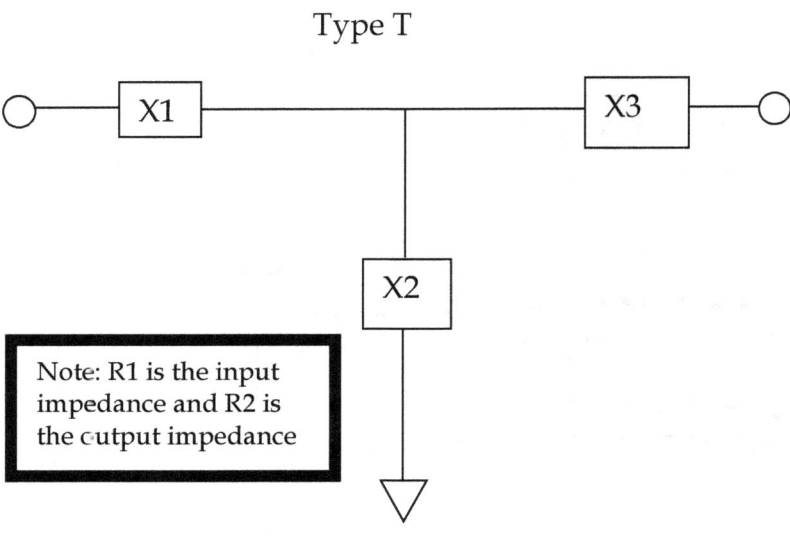

Note: R1 is the input impedance and R2 is the output impedance

Type Pi

The **transfer phase** is defined as the angle by which the current I2 **lags** the current I1. If Z1 and Z2 are **resistive** the phase is same as the phase lag of V2 with respect to V1. The phase angles of type A and type B sections are **dependent**, assuming that R1 and R2 are **independent**. The phase is also independent for the T and Pi sections within a range of + 180 Degrees to -180 Degrees (except zero). The design equations for these networks are shown below. Also, note that since the phase angle for the L sections is not independent it **cannot be chosen.** It will be what it will be, depending on R1 and R2. On the other hand the phase angle for the T and Pi networks **can be chosen** (or specified) as mentioned above.

The relationships for a *real source and real load* are shown below for the T, Pi and L networks.

T network:

$$X1 = \frac{\sqrt{R1R2} - R1\cos\beta}{\sin\beta} \qquad (3)$$

$$X2 = \frac{-\sqrt{R1R2}}{\sin\beta} \qquad (4)$$

$$X3 = \frac{\sqrt{R1R2} - R2\cos\beta}{\sin\beta} \qquad (5)$$

Pi network:

$$X1 = \frac{R1R2\sin\beta}{R2\cos\beta - \sqrt{R1R2}} \quad (6)$$

$$X2 = \frac{R1R2\sin\beta}{\sqrt{R1R2}} \quad (7)$$

$$X3 = \frac{R1R2\sin\beta}{R1\cos\beta - \sqrt{R1R2}} \quad (8)$$

Note: In these networks the phase angle β can be chosen from 180 degrees to -180 degrees except 0 degrees.

Type A network:

$$X1 = \frac{\sqrt{R1R2} - R1\cos\beta}{\sin\beta} \quad (9)$$

$$X2 = \frac{-\sqrt{R1R2}}{\sin\beta} \quad (10)$$

Also the phase angle is defined by:

$$\beta = \pm\cos^{-1}\sqrt{\frac{R1}{R2}} = \tan^{-1}\sqrt{\frac{R2}{R1} - 1.0} \quad (11)$$

Type B network:

$$X1 = \frac{R1R2\sin\beta}{R2\cos\beta - \sqrt{R1R2}} \quad (12)$$

$$X2 = \frac{R1R2\sin\beta}{\sqrt{R1R2}} \quad (13)$$

X3 = ∞. (14)

The phase angle is defined by:

$$\beta = \pm \cos^{-1} \sqrt{\frac{R2}{R1}} = \tan^{-1} \sqrt{\frac{R1}{R2} - 1.0} \qquad (14.1)$$

3.3　The Q matching technique.

Another approach to L section matching is the Q matching technique. This is explained below:

Two resistive terminations, one at the input to a network, and the other at its output can be simultaneously matched by adding two reactive elements between them. Calling the terminations Rlow and Rhigh the following can be done.

1) **Add a series element next to Rlow and a parallel one next to Rhigh.**
2) **The series element can be a capacitor or an inductor.**
3) **The parallel element has to be of the opposite type.**
4) **A series inductor with a parallel capacitor is a low pass circuit**
5) **A series capacitor with a parallel inductor is a high pass circuit.**

Figures 1 through 4 below show the configurations

Practical Impedance Matching Techniques

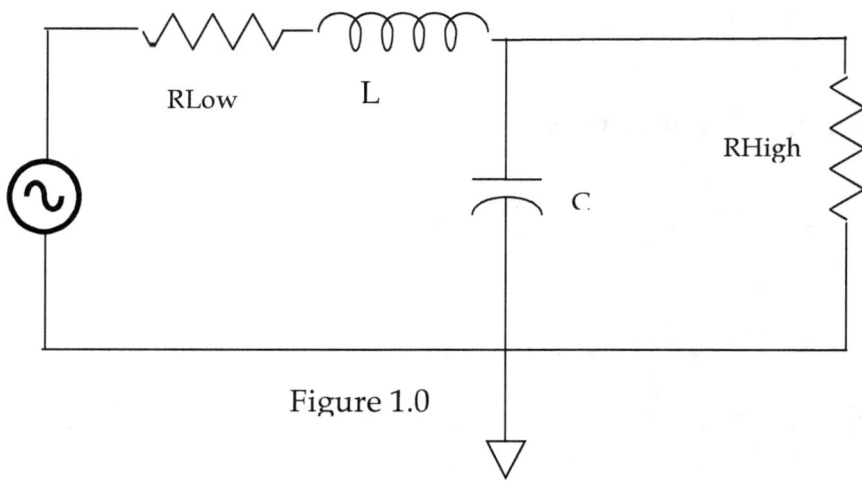

Figure 1.0

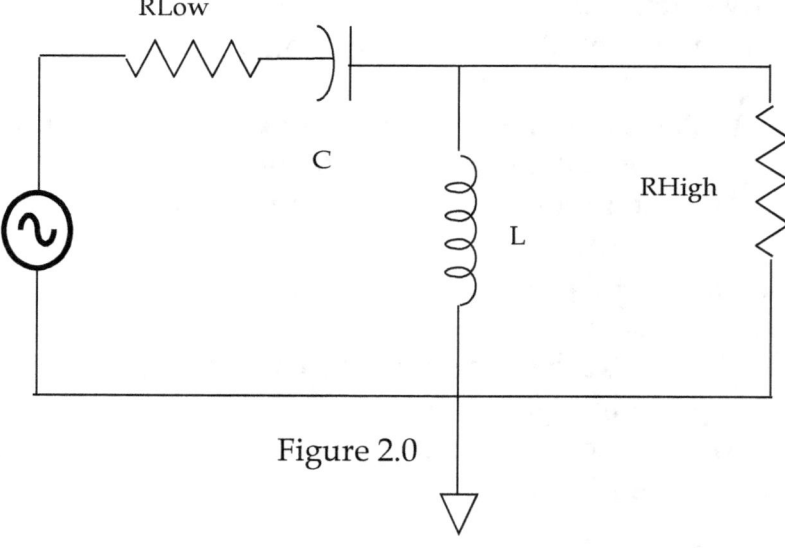

Figure 2.0

Practical Impedance Matching Techniques

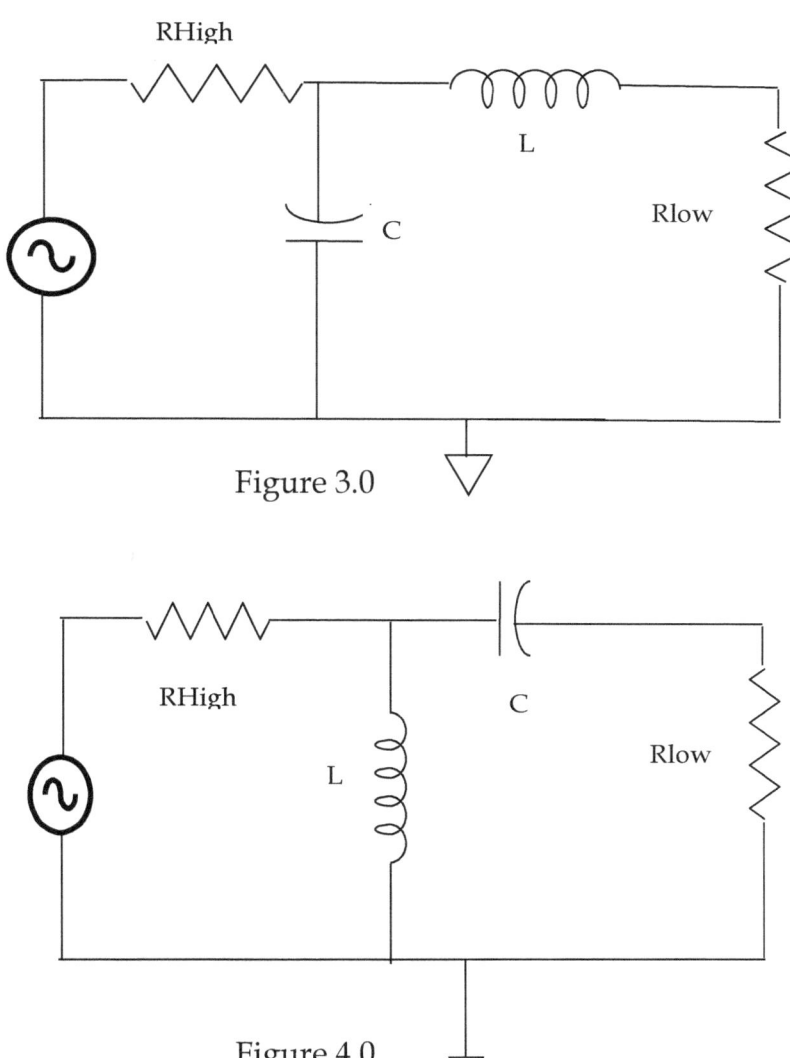

Figure 3.0

Figure 4.0

By examination of the circuits above, it can be seen that there are *two subnetworks* in each of the configurations. There is one series subnetwork and one parallel subnetwork, associated with the respective reactances.

By the law of matching, *these subnetworks must be complex conjugates of each other at the frequency of interest.* The Q factors of these subnetworks must be equal at the reference frequency. In other words:

where,
$$Q_s = Q_p \quad (15)$$

$$Q_s = X_s/R_{low} \quad (16)$$

$$Q_p = R_{high}/X_p \quad (17)$$

Figure 5 shows an example.

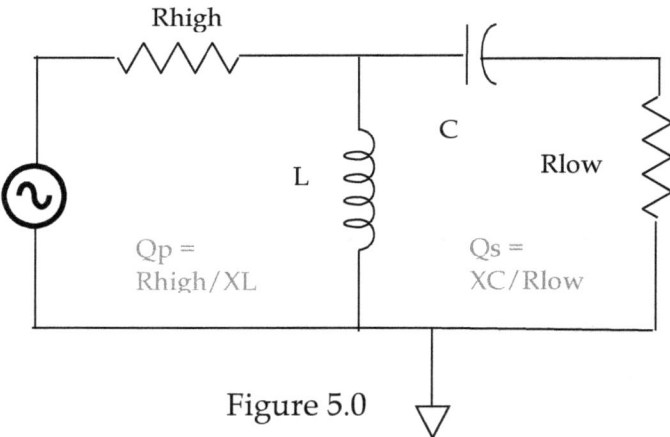

Figure 5.0

In order to find the expressions for use in this technique, we need to find *the equivalency of series and parallel forms.*

Lets assume a series circuit and first write the expression for Q for it:

$$Q = \frac{Xseries}{Rseries} \qquad (18)$$

Then,

$$Xseries = QRseries \qquad (19)$$

The terminal impedance for the series circuit now becomes:

$$Zseries = Rseries + jXseries = Rseries + jQRseries \qquad (20)$$

The conversion of the series impedance into parallel admittance is:

$$Ypar = \frac{1}{Zseries} = \frac{1}{Rseries(1+jQ)} \qquad (21)$$

Separating the real and the imaginary parts:

$$Ypar = \frac{1}{Rseries(1+jQ)} \frac{Rseries(1-jQ)}{Rseries(1-jQ)} \qquad (22)$$

$$Ypar = \frac{Rseries(1-jQ)}{Rseries^2(1+Q^2)} \qquad (23)$$

$$Ypar = \frac{Rseries}{Rseries^2(1+Q^2)} - \frac{jQ.Rseries}{Rseries^2(1+Q^2)} \qquad (24)$$

$$Y_{par} = \frac{R_{series}}{R_{series}^2(1+Q^2)} - \frac{j}{X_{series}\frac{1}{Q^2}(1+Q^2)} \qquad (25)$$

$$Y_{par} = \frac{1}{R_{par}} - \frac{j}{X_{par}} \qquad (26)$$

If we now equate the real and imaginary terms of the series and parallel circuit expressions we get:

$$\frac{1}{R_{par}} = \frac{1}{R_{series}(1+Q^2)} \qquad (27)$$

or,

R_{par} $= R_{series}(1+Q^2) \qquad (28)$

And,

X_{par} $= \frac{X_{series}}{Q^2}.(1+Q^2) \qquad (29)$

Further, the Q is:

$$Q = \sqrt{\frac{R_{par}}{R_{series}} - 1} \qquad (30)$$

From this result the two sub network Q's can be expressed as:

$Q_s = Q_p$ $= \sqrt{\frac{R_{par}}{R_{series}} - 1} \qquad (31)$

Of course this can be also expressed in normalized form by dividing the resistor values with the reference resistance (usually 50 Ohms)

Note: Rpar = Rhigh and Rseries = Rlow.

Once the Q is known we can calculate Xs and Xp as:

$$Qs = Xs/Rlow$$

$$Qp = Rhigh/Xp$$

4 Impedance matching of complex terminations.

If the terminations to be matched are complex. Then a modification of the above technique is used. It is best explained using an example. Let the source impedance be ZS=20 – j10 ohms and the load impedance be ZL = 6+j12 ohms. This is illustrated below in Figure L1.0.

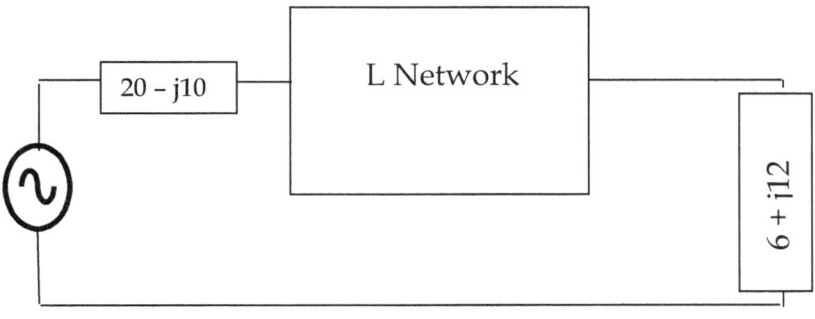

Figure L1.0

Converting the input impedance from a series configuration to a parallel one using the theory above as follows:

Practical Impedance Matching Techniques

$$Q = \frac{10}{20} = 0.5 \qquad \text{(LC1.0)}$$

$$Q^2 = 0.25 \qquad \text{(LC2.0)}$$

Then,

$$\text{Rpar} = 1.25 * 20 = 25.0 \qquad \text{(LC3.0)}$$

and,

$$\text{Xpar} = \frac{Xseries}{Q^2} \cdot 1 + Q^2 \qquad \text{(LC4.0)}$$

or,

$$\text{Xpar} = 50 \qquad \text{(LC5.0)}$$

So the input impedance is transformed to

$$ZS = 25 \| -j50. \qquad \text{(LC6.0)}$$

Here the sign $\|$ stands for "in parallel with".

A L - section can be used to match the 6 Ohms at the output with the 25 Ohms at the input. Using a type B L – section we get the following circuit. (Figure L2.0)

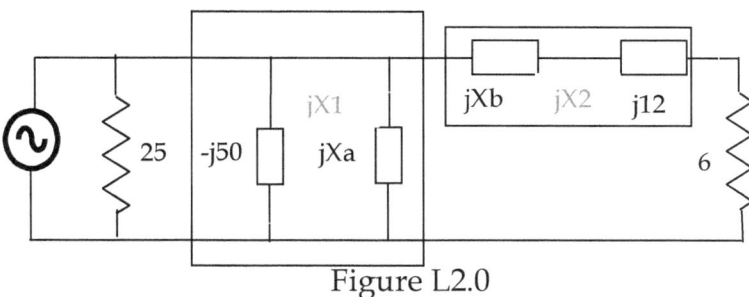

Figure L2.0

The type B matching network reactance, jXa and jXb are shown. If the matching scripts for the type B network are used then the following values are found:

$$X1 = \pm 14.05$$
$$X2 = \pm 10.68$$

Note that the phase angles define the sign of the reactance and therefore the type of component (L or C). The input reactance can be combined into a new reactance and from that the value of Xa can be found to be:

$$Xa = -19.54 \text{ or } 10.97$$

Similarly the series reactance at the output can be combined and the value of Xb can be found to be:

$$Xb = -1.32 \text{ or } -22.68$$

The dual values are a result of choosing the positive or negative phase angles.

If a type A network is to be used then the output impedance needs to be converted to a parallel form (the reason is that in a type A network the output load is in shunt). This leads to the circuit shown below in Figure L3.0

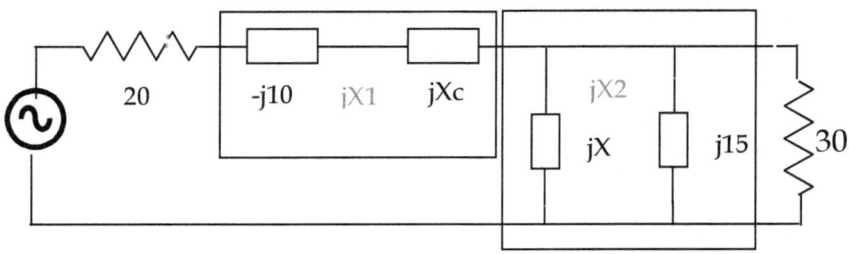

Figure L3.0

Following a similar procedure as in the case of a type B network, we get:

$$X1 = \pm 42.43$$
$$X2 = \pm 14.14$$
$$Xd = -23.20 \text{ or } -11.08$$
$$Xc = -4.14 \text{ or } 24.14$$
$$Xin = 10$$
$$Rin = 20.$$

It should also be noted in passing, that there is no reason to assume that all L – Section solutions must exist

In general, impedance matching of complex terminations is achieved by transforming one termination to be the complex conjugate of the other. Other related matching techniques are presented below.

If the load or source has imaginary parts then these can be absorbed in the matching network or eliminated by resonance as shown below.

In order to further the understanding of impedance matching the concept of the *quality factor Q and nodal Q* should be understood.

The unloaded Q of a reactive component is defined by:

Qu = Energy stored in the component/Energy dissipated in the component.

When the component is being used in a circuit, a quantity called the loaded Q is defined as follows:

Q_L = Energy stored in the component /Energy dissipated in the component *and the external circuit.*

The nodal Q of a L section matching network is defined as follows. At every node of a L section matching network there is a series impedance Rs + jXs. The nodal Q factor is then defined by:

$Q_N = |X_s|/R_s$.

The nodal Q is also computed as:

$$Q_N = \sqrt{[(R_{high}/R_{low})-1]} \qquad (32)$$

A. *Absorb the parasitics of the terminations:*

If the Q of the termination is less than the nodal Q (see definition above), then the reactance or susceptance may be absorbed into the matching network. For example consider the circuit below.

Practical Impedance Matching Techniques

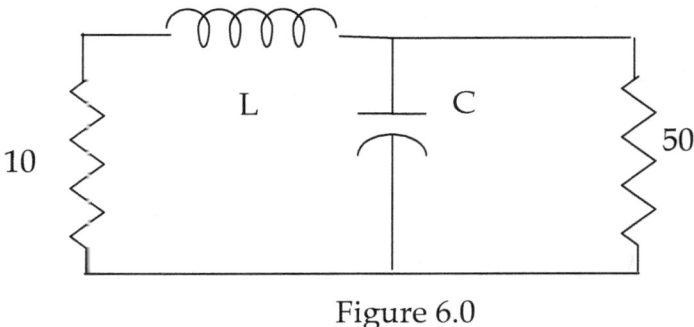

Figure 6.0

In figure 6.0 the nodal Q is:

$$Q_N = 2.0 \text{ (using eqn 32)}$$

(Please refer to figure 7.0) The total inductor in series with the input cannot exceed 3.18nH. If it is more than that, the Q will exceed 2.0 (the nodal Q).
Similarly the capacitance at the output cannot exceed 6.32pF. (This means the total shunt parasitic including Cpara). These limits are set by the ratio of the load resistor and the source resistor by definition of the nodal Q.

As long as the nodal Q value is not exceeded at the output or input the parasitics can be absorbed into the matching circuit. *Note however, that the absorption cannot be on both sides.* When the parasitic inductance *or* the parasitic capacitance is absorbed, the frequency response is unchanged.

Practical Impedance Matching Techniques

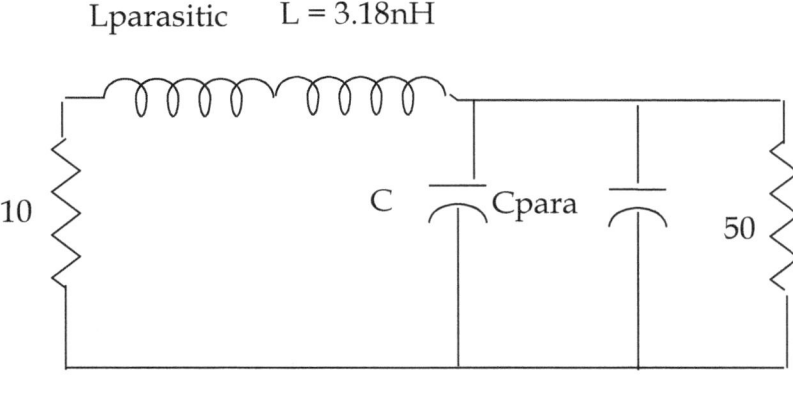

Figure 7.0

B Resonate excessive parasitic L or C.

If the parasitic L or C exceeds that maximum allowed value in the absorption technique then a couple of options open up for matching (*at one frequency*).

(B.1) Referring to figure 8.0 below, the capacitance Cmax, that exceeds the maximum allowable capacitance is resonated out by the parallel inductor, LR. This then, leaves the resistive part only visible for matching.

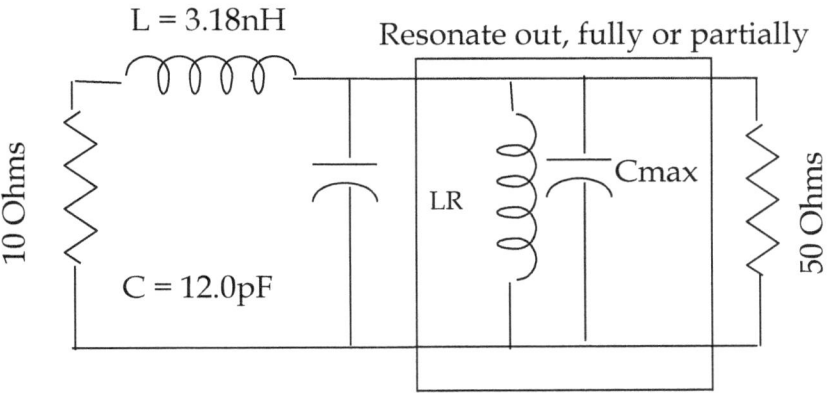

Figure 8.0

The second option is to resonate out a smaller portion of the capacitance Cmax and use the rest to perform the matching,

3.5   Bandwidth considerations:

In all these matching circuits and techniques, bandwidth has not been addressed. Either the matching is at one frequency or over a narrow band of frequencies. In this section the bandwidth is considered as a matching parameter and some techniques are presented for bandwidth control.

Increasing bandwidth: The simplest way to increase the bandwidth of the matching circuitry is to use a two section matching circuit. In other words take the matching L section and break it up into two sections. This is shown below. Choose an impedance Rmid = $\sqrt{(RS * RL)}$

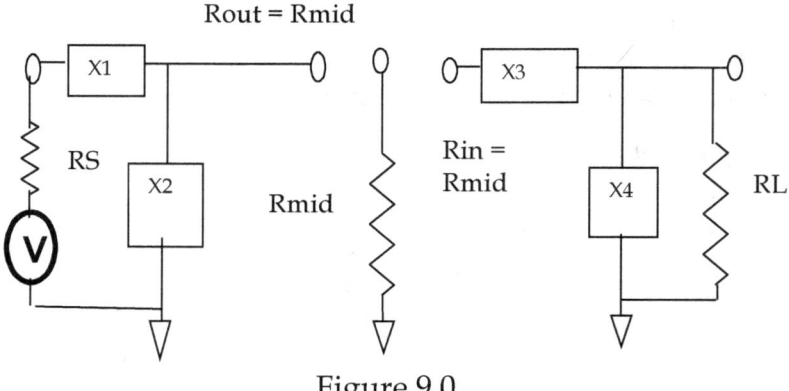

Figure 9.0

Then transform the input impedance to Rmid from RS and transform the output impedance RL to Rmid. This causes the reduction of the termination ratio and the nodal Q for both sections. The net result is an increase in bandwidth.

Optimally Rmid should be between RS and RL geometrically or,

$$RS/Rmid = Rmid/RL \qquad (33)$$

A two section matching network actually causes a large change in the bandwidth. However, this does not apply linearly, i.e. a three section network may not bring the same degree of change and subsequent networks bring less and less relative change. *After an increase to about 6 sections there is no noticeable change.* So the practical limit appears to be about 5 sections maximum.

To generalize the intermediate resistance levels we can formulate,

$$Ri = RS^{(k-i)/k} RL^{i/k} \qquad (34)$$

Here i = number of matching sections and k = interstage index counted from the source side. Again, a practical limit is i = 5.

Note that nodal Q's determine the bandwidth. A high nodal Q will lead to a narrow bandwidth while a low nodal Q will result in a broader bandwidth. From earlier considerations the nodal Q is determined by the *resistance ratio*. Therefore a high resistance ratio will generate a small bandwidth and vice versa. The resistance Rmid is *not a real device*, it is a contrivance to calculate the resistances required only. In reality the input impedance of the second stage represents the load equal to Rmid for the first section and so on. The output impedance of the first section is also Rmid. This acts as the source impedance for the second section. Calculations are based on these considerations.

Decreasing the bandwidth: What if the goal is a decrease in bandwidth? i.e. we *do not* want a circuit generating spurious responses outside the bandwidth. In this case a similar strategy as above is followed. Again a mid level impedance is used, Rmid. However, in this case Rmid is chosen *outside the range of RS and RL.*

The chosen Rmid is dependent on the values of RS and RL in the sense that Rmid can be chosen smaller than or larger than the range of resistances in the circuit.

If the resistances in question are low (say 100 Ohm match to 50 Ohm) then obviously choosing a higher resistance (Rmid) can cause problems so the choice should be to go for a lower (< 50 ohm) Rmid. If the range of matching is 15 ohms to 50 ohms then a choice of Rmid greater than 50 Ohms is a better choice since matching with low impedance levels (a few ohms) might be difficult. To show these strategies, consider the two circuits presented below.

Practical Impedance Matching Techniques

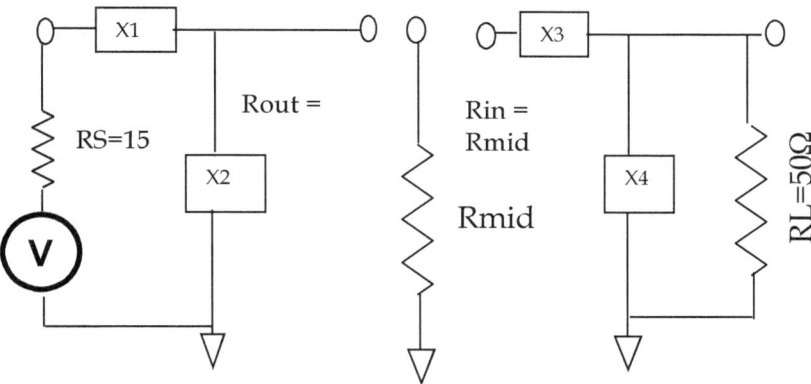

Figure 10.0 In this circuit Rmid is chosen lower than 15 Ohms.

In Figure 10.0, Rmid is chosen below 15 Ohms. Why? In order to answer this question it is best to use an example. Figure 11.0 shows a matching circuit using a low pass first stage and a high pass second stage.

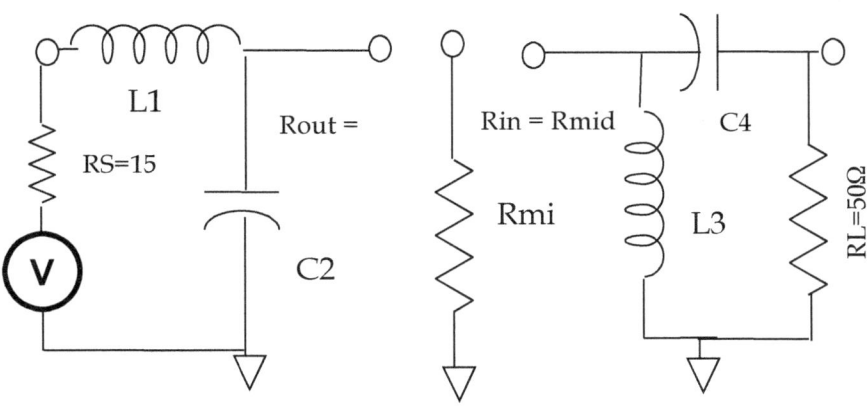

Figure 11.0

There appears to be no analytical way of determining Rmid. It has to be chosen. In the case of Figure 11.0 a value of 250 Ohms is chosen. Then:

The nodal Q for the first LC circuit is 15.7. For the second circuit it is 4. Obviously the bandwidth will be governed by the larger of the two. Then L1 and C2 can be determined by the methods described before.

A dual of this circuit is the high pass - low pass combination. The treatment is the same.

3.6   Using RF transformers and baluns for matching

RF transformers and baluns are a very simple way of matching impedances in circuits. RF transformers and baluns are related components with slight differences in operation. In fact RF transformers can be used as baluns in some cases. This section of the book deals with these helpful components.

RF transformers and baluns form a very useful category of electromagnetic (EM) devices that are used in a number of applications. These are impedance matching, voltage and current conversion tools that most electronic engineers should know about.

In the RF/Microwave world of design, these devices sometimes become indispensable because of the functions mentioned above. The transformer or the balun can do very efficiently and cost effectively, that which would be prohibitive in cost and effort if done on a piece of a semiconductor substrate using active and passive devices.

For these reasons the usage of RF transformers and baluns has continued unabated over the years and one can either buy, construct or design these EM devices on a board or a semiconductor chip.

A number of CAD tools exist today that can help to do this quite satisfactorily, and the author enthusiastically recommends those. A web search can list a multitude of these tools both free and for purchase.

This book is limited to understanding the basics of these devices and their usage in most applications without going in depth on the design. That will have to wait for a subsequent book!

The good news is that if you need to use some of these devices on your board level application you can easily purchase the appropriate device from a number of well known vendors. The author has used devices from these vendors with success and this can be construed as a recommendation.

The RF Transformer: the absolute basics.

It behooves us to understand the basics about transformers for a further study of RF transformers. This chapter explores the basics of transformers in general.

I.1   Transformer symbol

Figure 1.0 shows the generally accepted transformer symbols.

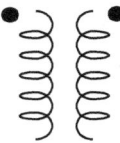

(a) Air Core transformer

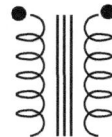

(b) Iron Core transformer

Figure 1.0

The air core transformer has no solid core and may be built using windings on an insulated tube or cylinder or with no core at all! Air core transformers provide the least coupling between the coils but are ideal for high frequencies. Permeability is = 1.0

The iron core transformer has a metallic core that may be made of a solid mass of metal or a laminated core. Much better coupling is available. The permeability is = 500+. An additional type of core is ferrite that provides great coupling and permeabilities of 500 to 9000 or more. Ferrite cores are also classified as "soft" cores. These cores are not suitable for low frequency operation.

I.2   Transformer dot convention or rule

The dots placed on the symbols of the transformers signify the phase relationships of the coils of the transformer.

If the currents are both entering or leaving the dotted terminals, the mutual fluxes will reinforce each other. Or defined in another way, if the voltage at the dotted end of the primary is positive with respect to the undotted end, then the dotted end of the secondary is also positive with respect to the undotted end.

In addition, if the primary current flows *into* the dotted end of the primary winding then the secondary current also flows *out* of the dotted end of the secondary.

I.3   Basic Transformer equations

The following simple equations can be used to understand transformer behavior.

$$n = N_2/N_1 \qquad \text{I.3 (1)}$$
$$V_2 = nV_1 \qquad \text{I.3 (2)}$$
$$I_2 = I_1/n \qquad \text{I.3 (3)}$$
$$Z_2 = n^2 \cdot Z_1 \qquad \text{I.3 (4)}$$
$$L_s/L_p = n^2 \qquad \text{I.3 (5)}$$

Here:

n = turns ratio of the two windings

V_2, V_1 = secondary and primary voltages between the respective terminals.

I_2, I_1 = Primary and secondary currents.

Z_2, Z_1 = secondary and primary impedances.

L_s, L_p = secondary and primary inductances.

I.4 Basic Transformer circuit model

So far the discussion about the transformer has been based on an ideal transformer. However, no transformer is ideal.

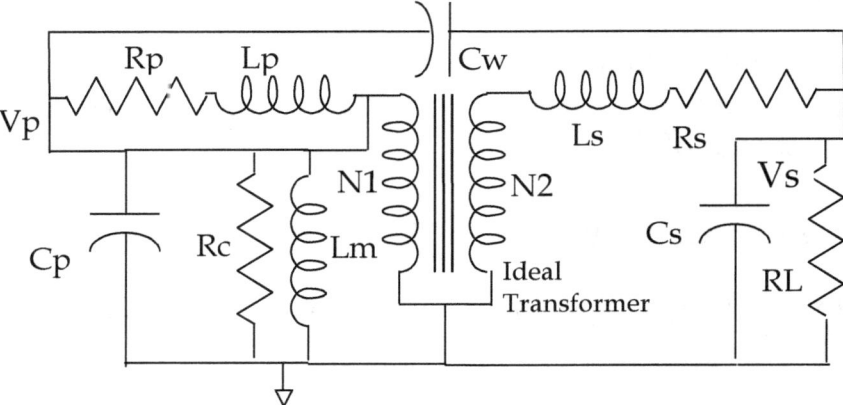

The components of the practical transformer that cause the non ideality, are the ones shown outside the ideal transformer symbol. These effects will be explored below.

I.4.1 Rp and Rs:
These are the primary and secondary winding resistances.

I.4.2 Lp and Ls: These are the primary and secondary leakage inductances. These inductances represent the primary and secondary flux that does not flow through the core but is lost or leaks out. This is a loss in the system.

I4.4 Rc: This resistance models the hysterisis and eddy current losses within the core (core loss) owing to the ac flux in the core.

I4.5   Lm: This is the magnetizing inductance, an inductance associated with the magnetization of the core, i.e that which establishes the flux in the core.

I4.6   Cp and Cs: These are the lumped capacitances of the primary and the secondary circuits.

I4.7   Cw: This is the interwinding lumped capacitance.

I4.8   RL: This represents the load resistance.

Depending on the construction and materials of the transformer, these parameters will vary and therefore the transformer's performance will vary.

Transformer suppliers should be able to supply these parameters for their products or they can be derived from parameters that *are* specified.

Once the parameters are available, a simple simulation may be done to generate an approximation to the final performance of the transformer. It must be noted that the transformer is a complex device and final performance has to be measured from an actual sample of the device to confirm that data.

I5.0   Transformer insertion loss:

In general insertion loss is defined as the loss in signal power when a device is inserted in the transmission path of a signal on a transmission line or circuit.

Insertion loss is usually stated in decibels (dB). So if the transmitted power is PT and the received power at the load, after the insertion of the device is PR, then the insertion loss is defined as:

$$IL\ (dB) = 10Log(PT/PR)$$

Where the Log is to the base of 10.

Insertion loss is also defined for filters as the ratio of the signal level in a test configuration *without* the filter installed, to the signal level *with* the filter installed. So if the signal level without the filter is

V1 and the signal level with the filter installed is V2 then the insertion loss in dB is:

$$20Log(|V1|/|V2|)\ dB$$

If using scattering parameters use the following expression:

$$10Log(|s21|^2/1 - |s11|^2)$$

where the symbols have their usual meaning.

Transformers also have an insertion loss specification. This is a figure of merit for a RF transformer.

The low end (or low frequency) loss is determined by the primary inductance. The high frequency insertion loss is dependent on the losses in the interwinding capacitance and the series inductance.

In addition, for transformers with metallic cores, the permeability is directly proportional to the temperature of operation. As the temperature decreases, the permeability *decreases* which causes an *increase* in the insertion loss.

I.6 Amplitude and phase unbalance

When using center tapped secondary windings to provide differential signals at the output, the amplitude and phasing of the outputs, becomes an important specification for the center tapped secondary transformer configuration.

If the input signal to the primary is V1 and the two outputs are V2 and V3, then the amplitude unbalance in dB is defined as:

$$20 Log(|V2|/|V3|)$$

The phase unbalance is defined in degrees. This is the deviation in phase from the exact 180 degrees between the two outputs.

I.7 Configurations of RF Transformers

I.7.1 DC isolated center tapped secondary

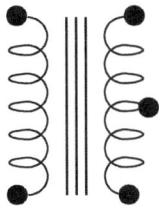

Figure 3.0

Typical frequencies of operation are 0.01 to 1400 Mhz.

Typical impedance ratios vary from 1.0 to 16.0. Unbalance is usually excellent. Typical power handling capability is upto 1.0 Watt. Applications include impedance conversion, baluns and unbalanced to balanced operation.

I.7.2 DC isolated center tapped primary and secondary

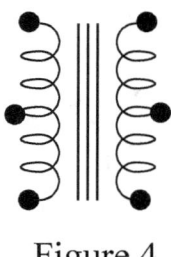

Figure 4

Typical frequencies of operation are from 0.004 to 500 Mhz.

Impedance ratios vary from 1.0 to 25. Unbalance is usually excellent. Typical power handling capability is up to 0.25 Watt. Applications include impedance conversion, baluns and balanced to balanced operation.

I.7.3 DC isolated primary and secondary

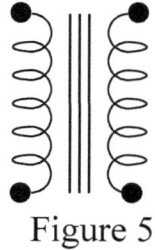

Figure 5

Typical frequencies of operation are 0.01 to1200 Mhz.

Impedance ratios vary from 1.0 to 36. Unbalance is usually average. Typical power handling capability is up to 0.25 Watt. Applications include impedance conversion, baluns and balanced to balanced operation.

I.7.4 Autotransformer

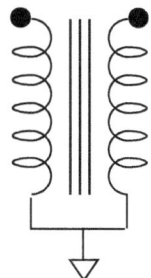

Figure 6.0

Typical frequencies of operation are from 0.05 to 2200 Mhz. Impedance ratios vary from 0.1 to 14. Typical power handling is up to 0.25 Watt. Applications include impedance conversion. There is no DC isolation between primary and secondary.

It can seen from the equations presented above, I.3 (1) through I.3(4), repeated here for convenience, that a transformer provides impedance matching as well as level conversion. The level conversion is achieved through the primary and secondary windings as is the impedance conversion. For example if the output impedance of an amplifier is 10 ohms and the load is 50 ohms then using the equations presented above,

$$n = N2/N1 \qquad \text{I.3 (1)}$$
$$V2 = nV1 \qquad \text{I.3 (2)}$$
$$I2 = I1/n \qquad \text{I.3 (3)}$$
$$Z2 = n^2 \cdot Z1 \qquad \text{I.3 (4)}$$

a transformer with a turns ratio of = 2.23 such that Z2 becomes 50 ohms while Z1 (the output impedance of the amplifier) is 10 ohms. Of course the output voltage level will also change to 2.23X the input voltage level. The output current level will change by 1/2.23.

Note also that the transformer can provide balun operation. For example a 1:1 RF transformer may be used to provide differential to single ended conversion by using the primary winding in the differential mode and grounding the appropriate terminal at the output and taking the output from the ungrounded output terminal.

Baluns

The term BALUN is an abbreviation for "<u>bal</u>anced – <u>un</u>balanced". In addition:

A balun can be viewed as a three port power splitter. The two outputs should be equal in amplitude and opposite in phase. This implies the outputs are shifted by 180° in phase. The time waveform of one balanced output is the negative of the other balanced output. Typically, the unbalanced input is matched to the characteristic impedance of the transmission line that feeds it. (usually 50 Ω). A balun can be used bi-directionally.

Note that a RF Transformer can be used as a balun.

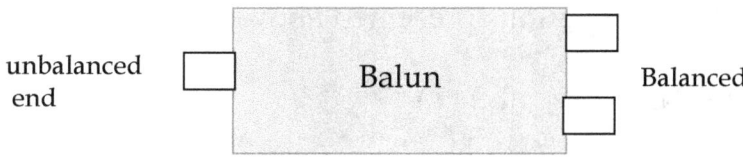

Figure BAL 1.0

Parameters:

1.0 The impedance ratio is simply:

$$Z_{unbalanced}/Z_{balanced} = \text{Turns ratio}^2$$

2) **Phase Balance**: How close the balanced outputs are to having equal power and 180° phase, is called balance. It is a critical parameter for baluns. Typical phase balance for standard microwave baluns is 15° max and 10° typical.

3) **Amplitude Balance:** Usually specified in dB and actually provides the match between output power magnitude.

4) **Common Mode Rejection Ratio:** If two identical signals with the same phase are applied to the balanced ports of the balun (called 'common mode' or 'even mode' signals), they will be either reflected or absorbed. The amount of attenuation this signal suffers from the balanced to unbalanced port is called common mode rejection ratio (CMRR) and is expressed in dB.

5) **Impedance Ratio/Turns Ratio:** The unbalanced impedance of a balun is matched to the input transmission line. However, the balanced impedance can take any value. The ratio of the unbalanced impedance to the balanced impedance is the impedance ratio, and is usually stated as 1 : n (e.g. 1:1, 1:2, 1:4).

Insertion and Return Loss: Low insertion loss and high return loss means more power available for succeeding functions in the cascade and a better dynamic range. Also it causes less distortion of signals in previous stages of the system

Balanced Port Isolation: Isolation has the same connotation as in other power dividers and couplers. The insertion loss from one balanced port to the other in dB is the isolation. Generally baluns *do not* offer high levels of isolation.
The reason is that the even mode is reflected instead of being properly terminated with a resistive load.

DC/Ground Isolation: DC isolation refers to whether the unbalanced port has a DC connection to one of the balanced ports. Ground isolation is whether there is a connection between the unbalanced ground and the balanced signals or grounds.

> *An excellent treatise on Balun basics can be found in: Balun Transformers, Magic-Ts, and 180° Hybrids By: Doug Jorgesen and Christopher Marki published by Marki Microwave.*

A simple Balun:

Please refer to the figure below that shows a simple LC balun.

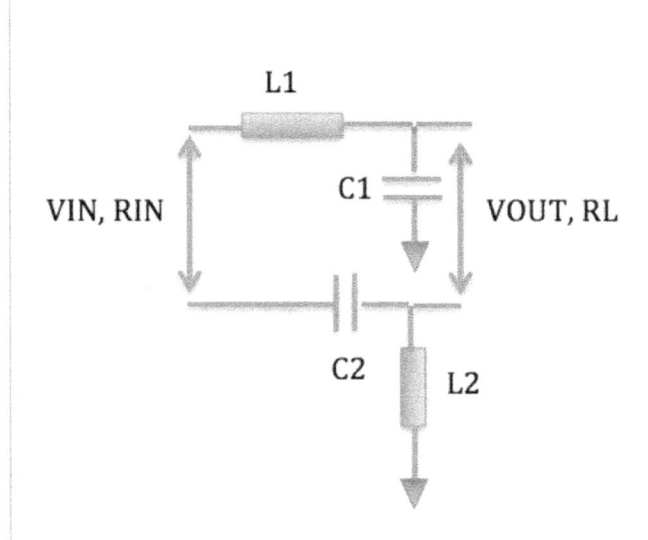

Description of operation:

One of the simplest baluns can be designed using L and C elements as shown:

The design equations have been embodied in a Javascript calculator The input of this circuit is a unbalanced waveform and the impedance is set by the values of the L/C combinations. The output is another balanced signal, with the required output impedance also calculated using the input and output impedances. This circuit may be implemented using discrete elements for use as a impedance transformer or a balun.

A javascript calculator is available for the calculation of the parameters of this balun and can be found, free of charge, in the Signal Processing Group Inc. blog at :

http://www.signalpro.biz/wordpress/lc-balun-calculator/

Transmission line baluns:

"The TLT transmits the energy from input to output by a transmission line mode and not by flux-linkages as in the conventional transformer.

As a result the TLT has much wider bandwidth and higher efficiencies than its conventional counterpart. With proper core materials and impedance levels of 100 ohms or less, bandwidths of about 100 MHz and efficiencies approaching 99% are possible today when matching 50 ohms up to 100 ohms and 50 ohms down to about 3 ohms. Future research and development, especially with impedance ratios less than 4:1, should make TLTs operate at much wider bandwidths." - **Jerry Sevick, Bell Laboratories (Retired) and Consultant** in *"A Simplified Analysis of the Broadband Transmission Line Transformer "* published in February 2004 in *High Frequency Electronics. Ref BAL1.*

In conventional transformers the inter-winding capacitance resonates with the leakage inductance producing a loss peak. This mechanism limits the high frequency response.

In transmission line transformers, the coils are so arranged that the inter-winding capacity is a component of the characteristic impedance of the line, and therefore forms no resonances which seriously limit the bandwidth.

There are a quite a few transmission line balun configurations for use under different conditions. A popular balun is the Guanella balun as shown below:

Guanella balun

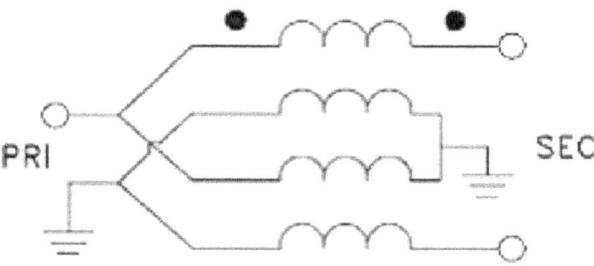

The following is an explanation of the operation of this balun.

The operation of the Guanella balun is not intuitive so some tutorial information is in order...

It is important to understand that the Guanella balun is not about flux-linkages and conventional-transformer action. *Those concepts have nothing to do with the Guanella (current) balun.*

The 1:4 Guanella (current)-balun is made from two 1:1 Guanella (current)-baluns. Except for winding direction, they should be as identical in construction as possible to minimize any variation in signal delays. Variations will increase signal loss at the higher frequencies.

The 1:1 Guanella (current)-balun can be made in multiple ways:
A) coiling coax.
B) Winding a transmission-line (coax or two-wire) on a toroid-core or rod. A ferrite core or rod is preferred.
C) Threading coax thru ferrite-beads.

Figure 7.0 below shows the usual schematic used for a 1:1 current-balun.

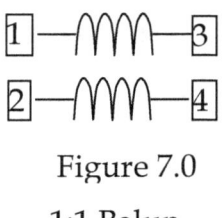

Figure 7.0

1:1 Balun

Note that this is just two identical transmission lines. To further underline this fact, the same 1:1 Balun is redrawn below in terms of simple transmission lines.

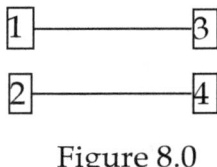

Figure 8.0

It is assumed that the transmission lines above do not cause signal loss so whatever is input between terminals 1 and 2 is output between terminals 3 and 4. In practice this will not be the case, but it can be approximated by using really low loss and high quality lines.

Let us now connect two 1:1 baluns as shown below.

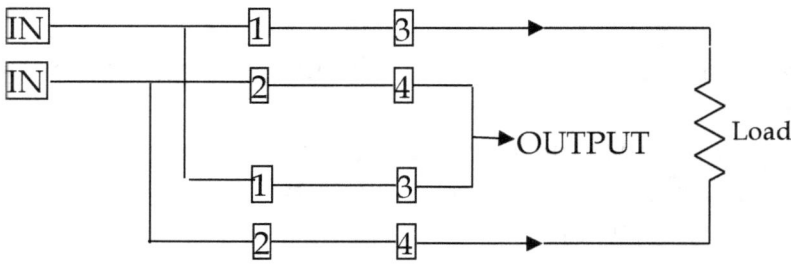

Figure 9.0 1:4 Balun

Note how the inputs are in *parallel* and the outputs are in *series*.

If the input signal is a voltage V, the input power to the balun is simply:

$$Pin = V^2 / ZIN. \qquad (1)$$

Here ZIN is the input impedance between the inputs.

The output power in the load is calculated as follows:

The input voltage at pins IN and IN is *V*. Therefore from the earlier treatment of the 1:1 balun the same voltage must appear across the output terminals of the two 1:1 baluns that form the above balun. However, the outputs are connected in series therefore the total voltage across the output terminals 3 and 4 must be *2V*.

The power across the load resistor then is simply:

$$Pout = 4V^2 / R_{load} \qquad (2)$$

Therefore since no power is lost in the baluns (as stipulated above). The power into the balun must be equal to the power at the output into the load. Then:

$$Pin = Pout$$
$$V^2 / ZIN = 4V^2 / R_{load} \qquad (3)$$

Or,

$$R_{load} / ZIN = 4 \qquad (4)$$

For example if ZIN = 50 then R_{load} = 200 would be the correct impedance at the output. There is a 1:4 impedance conversion using this balun.

The balun is a bilateral device. If a 50 Ohm impedance is applied at the output then the correct impedance at the input would be 12.5 Ohms or a 4:1 impedance conversion.

To use the Guanella balun properly the characteristic impedance Z_0 should be chosen correctly. The right value is to divide the high impedance side impedance by 2 for the correct Z_0 for the output lines.

Finally, the common node at the output can be grounded as shown above to get balanced outputs as needed.

Ruthroff balun:

Ratio of 1:1. Please refer to the figure below.

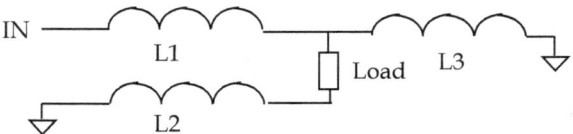

So the 1:1 Ruthroff balun looks very similar to a Guanella balun except for an extra winding that generates a voltage divider at the output.

It is apparent from the figure that windings L1 and L2 are the same as a Guanella balun. If the reactance of the winding L3 is 10 or more times the impedance of the load RL, the device acts just like a Guanella balun because its impedance is much higher than RL.

Ruthroff also has a 1:4 balun shown in the figures below.

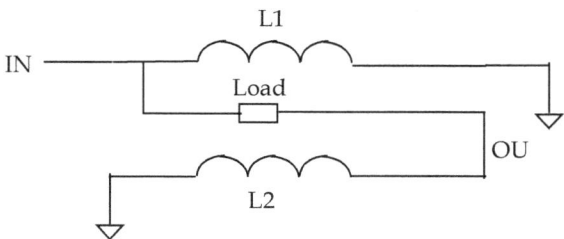

To read more about Ruthroff baluns and ununs please see the original paper : "**Some Broad-Band Transformers**" by L. RUTHROFF, MEMBER, IRE, *Proceedings of the IRE August 1959.*

In general Ruthroff baluns are lower frequency in performance. However, if the transmission lines are kept short then the Ruthroff designs become of practical use.

More information on these types of baluns is available in the excellent overview papers cited above by Sevick and Ruthroff and interested readers can do more in depth reading using those references.

These baluns are not restricted to be 4:1 only. They can be made with other ratios although they become more cumbersome and difficult with the increase in the ratios.

Marchand balun:

The Marchand balun is a good *broadband* balun and can be implemented using discrete components as well as integrated ones.

Practical Impedance Matching Techniques

The conventional Marchand balun employing two quarter wave-length coupled line sections is perhaps the most attractive topology due to its wideband performance.

The main disadvantages of the conventional Marchand balun is its large size increasing the fabrication cost for MMIC application
except at the highest mm-wave frequencies.

The figure below shows the conventional Marchand Balun.

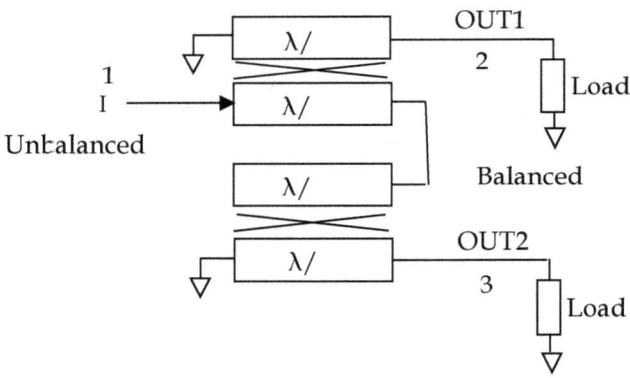

In terms of s parameters, the ideal Marchand balun provides:

$$s11 = 0 \qquad (M1.0)$$
$$s21 = -s31 \qquad (M2.$$

Even and odd mode impedance:

A brief digression from the main topic is necessary here to understand the concept of even and odd characteristic impedance of coupled lines.

The odd mode of impedance occurs when the coupled line is driven differentially while the even mode occurs when the coupled line is driven in common mode. i.e both line signals are in phase.

When the lines are driven in the odd mode, it appears that a virtual ground plane has been inserted between the traces. Although this is an imaginary plane its influence on the signals is as if it were real.

In the common mode or even case there will be no such creation of a virtual ground plane. The effect of the common mode effect is to increase the impedances slightly.

Let us call the even mode impedance Zoe and the odd mode impedance Zoo. Then the following holds true:

$Zo = (Zoo.Zoe)^{0.5}$ also additionally
$Z_0^2 = Z_{0e} * Z_{0o}$
a restatement of the above and,
$Z_{0e} = Z_0 * [\,(1+k)/(1-k)\,]^{0.5}$
$Z_{0o} = Z_0 * [\,(1-k)/(1+k)\,]^{0.5}$

where k is the coupling factor given by:

$k = 10^{-C(db)/20}$

Sometimes the coupling coefficient β is used where,

$$\beta = Zoe - Zoo/Zoe + Zoo.$$

In terms of a 2 port system we can also write:

Coupling = C = 10Log(P1/P3) dB where P1 is the input port applied power and P3 is power extracted at port 3 as shown below. So if C = 20dB, then the coupling factor is 0.1.

In terms of the coefficient β we can also write:

Coupling C = -20log β dB.

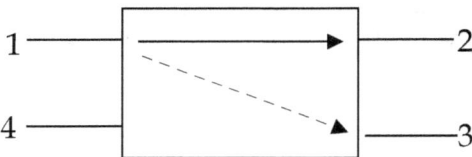

The web is not exactly replete with information on even and odd analysis but this brief explanation will serve our purposes for the subject in question.

When the Marchand balun is made using coupled lines there will be much lower restrictions on the value of Zoe. A marchand balun with good performance can be developed with Zoe ≈ 3 to 5 times larger than Zoo. That's it.

References:

1.0 "RF and microwave coupled line circuits", Mongia R.K, Bahl I.J, Bhartia P. and Hong J.
2.0 "Analysis and design of lumped element Marchand baluns", Johanson, T.K and Kroger V. 17th International

conference on microwaves, radar and wireless communications, 2008.

3.0 "New classes of miniaturized planar Marchand baluns", Fathelbab W.M, Steer M.B. IEEE transactions on microwave theory and techniques, vol 53, No. 4, April 2005.

Part IV

Transmission line matching circuits:

4.1 Simple cascaded line matching:

At elevated frequencies, transmission line matching becomes practical. The simplest match with a transmission line is given by adding a length of line with characteristic impedance Zo, between the resistive load and resistive source.

Characteristic impedance of the line
Zo = √(RL.RS). (35)
The electrical length is:
θ = 90 Degrees (Note electrical lengths of 270 Degrees, 450 Degrees etc are also a solution)

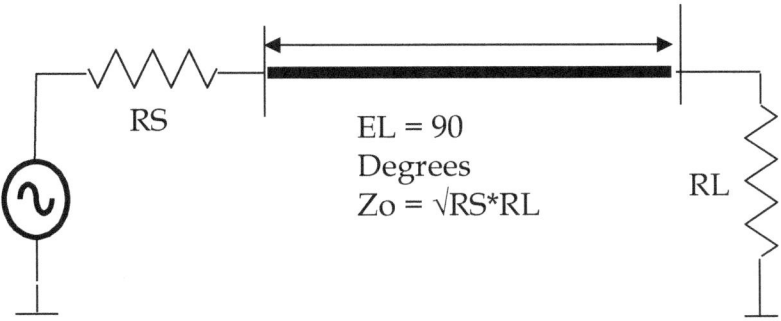

4.1.1 Definition of Electrical length:

As a note, the electrical length of a transmission line is defined by the following:

1.0 The wave number or phase constant = $\beta = 2\pi/\lambda$

2.0 The electrical length is defined by $\theta = \beta l$ where l = physical length

3.0 $\theta = \beta l = (l/\lambda) * 360$ degrees

Here λ is the wavelength of the signal in the applicable dielectric sometimes called the guide wavelength.

4.0 For frequencies in Ghz, [360 * fGhz * l(cm) * $\sqrt{\varepsilon eff}$]/30 cm

In this case frequency is in Ghz, physical length is in centimeters.

For example:

Let frequency be 1 Ghz.
Let $\lambda = 0.8\ \lambda$(air) or $\sqrt{\varepsilon eff} = 1.25$
Let $l = 0.1$ meters = 0.1E2 centimeters

Then :

$\theta = [360 * 1 * 0.1E2 * 1.25]/30$ degrees

$\theta = 150$ degrees

4.1.2 **Definition of β:** Sometimes β is referred to as the phase constant of the line or guide. If the Cartesian coordinate system is used and a coordinate, say "z" is used as the direction of wave propagation then βz measures the instantaneous phase at point z on the line with respect to z = 0.

In addition, voltage or current on the line is the same at any two points separated in z such that βz differs by multiples of 2π. Since the shortest distance between points where voltage or current is at the same phase is a *wavelength*, then:

$\beta\lambda = 2\pi$

(replacing z by λ),

$\beta = 2\pi/\lambda$

Complex impedances can also be matched with a section of transmission line. The only stipulation is that RS should not be equal to RL. If this is true then the characteristic impedance of the line is given by:

$$\sqrt{\{[(RS^2 + XS^2)RL - (RL^2 + XL^2)RS]/(RS - RL)\}}$$

(36)

where, the source impedance is RS+jXS and the load impedance is

RL+jXL.

and the electrical length is:

$\theta = \tan^{-1} [Zo(RL-RS)/(XSRL - XLRS)]$

4.2 The quarterwave transformer:

The quarter wave transformer is a quarter wavelength of transmission line with a characteristic impedance of Zi placed between a transmission line of characteristic impedance Zo and a real load impedance of RL. Figure 12.0 below depicts the circuit.

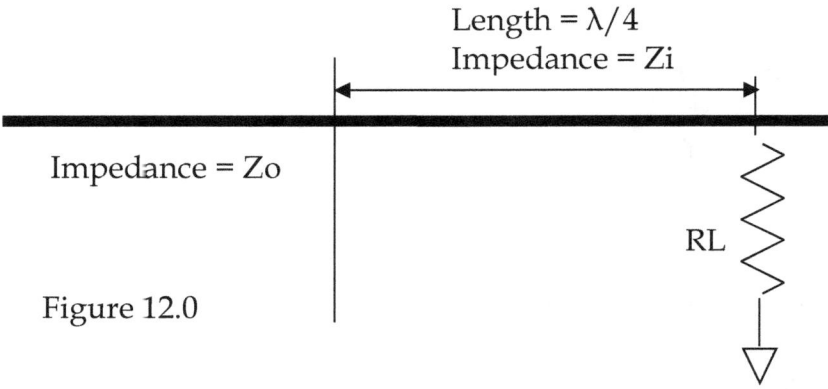

Figure 12.0

Obviously the impedance is dependent on the width of line and the thickness of the insulation on which it resides (if it is a microstrip line). In any case, transmission lines can be any style. The characteristic impedance can usually be calculated fairly easily.

The analytic expressions associated with this type of matching network are presented below.

Practical Impedance Matching Techniques

The input impedance of a transmission line terminated in a load ZL is given by:

$$Zin = Z1 \frac{ZL + jZ1\tanh(\gamma l)}{Z1 + jZL\tanh(\gamma l)} \qquad (38)$$

Here,

$Z1$ = characteristic impedance of the line

ZL = load
γ = propagation constant

If the load is real and equal to RL,

Then,
$$Zin = Z1 \frac{RL + jZ1\tanh(\gamma l)}{Z1 + jRL\tanh(\gamma l)} \qquad (39)$$

If the length of the transmission line is a *quarter wavelength*, then the input impedance is purely real.

Then,

$$\gamma l = \frac{2\pi}{\lambda} \cdot \frac{\lambda}{4} = \frac{\pi}{2} \qquad (40)$$

λ = wavelength of the signal

The terms tanh() become unbounded. Taking the limit we get:

$$Zin = \frac{Z1^2}{RL} \qquad (41)$$

If the system is matched then the input impedance must be equal to the characteristic impedance Zo. Or,

$$Zin = \frac{Z1^2}{RL} = Zo \qquad (42)$$

Which leads to:

$$Z1 = \sqrt{(ZoRL)} \qquad (43)$$

For example: Match a 50 ohm line to a 100 ohm load. Determine the characteristic impedance of the matching quarterwave line.

Here RL = 100.
Zo = 50. Then:
Z1 = √(100*50) = 70.71 Ohms.

This is one of the simplest ways to match a load to the line.

4.3 Frequency response of the quarter wave transformer.

The frequency response of the reflection coefficient is given by:

$$\Gamma(\omega) = \frac{Zin(\omega) - Zo}{Zin(\omega) + Zo} \qquad \text{f1.0}$$

$$Zin(\omega) = Z1 \frac{ZL + jZ1\tanh(\beta(\omega)l)}{Z1 + jZL\tanh(\beta(\omega)l)} \qquad \text{f2.0}$$

$$Z1 = \sqrt{(ZoZL)} \qquad \text{f3.0}$$

Practical Impedance Matching Techniques

$$\beta(\omega)l = \frac{2\pi}{\lambda} \cdot \frac{\lambda o}{4} = \frac{\pi}{2} \quad (\frac{f}{fo}) \qquad \text{f4.0}$$

f = frequency variable,
fo = reference frequency, design center frequency for the quarter wave transformer.

4.4 Multisection matching transformer:

4.41 The binomial transformer

A multisection matching transformer is made by connecting K transmission lines in series between the feeder line of characteristic impedance Zo and the load ZL.

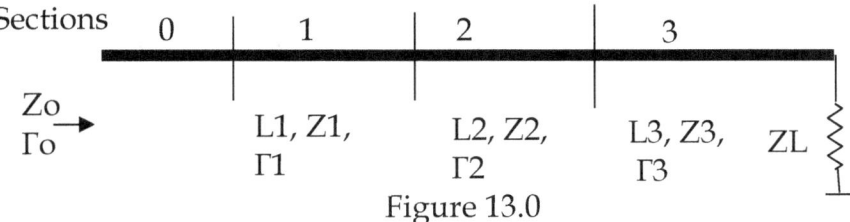

Figure 13.0

Where , Li = line lengths, Zi = impedances, Γi = reflection coefficients for each section of line.

Note that the reflection coefficient is defined by,

$$\Gamma = \frac{Z - Zo}{Z + Zo} \qquad (44)$$

where Zo is the source impedance and Z is the load impedance.

Thus for section 0 above we have a reflection coefficient of:

$$\Gamma o = \frac{Z1 - Zo}{Z1 + Zo} \quad (45)$$

for any n except N we have:

$$\Gamma n = \frac{Z_{n+1} - Z_n}{Z_{n+1} + Z_n} \quad [1 \leq n \leq N-1] \quad (47)$$

and for the last section:

$$\Gamma N = \frac{ZL - ZN}{ZL + ZN} \quad (48)$$

We now make each section of the line the *same length*. In this case the reflection coefficients are of the same sign. Under these assumptions the total reflection coefficient may be written as:

$$\Gamma(\theta) = \Gamma o + \Gamma 1 e^{-2j\theta} + \Gamma 2 e^{-j4\theta} + \ldots + \Gamma N e^{-j2N\theta} \quad (49)$$

An additional assumption made at this stage is that the reflection coefficients are symmetric. This implies that $\Gamma o = \Gamma N$, $\Gamma 1 = \Gamma N-1$ and so on.

Equation (49) leads to the following summation:

$$\Gamma(\theta) = A \sum_{n=0}^{n=N} Cne^{-j2n\theta} \quad (50)$$

$$\theta = \frac{\pi}{2} \frac{f}{fo} \quad (50.1)$$

Cn is the binomial coefficient and A is the amplitude factor. The binomial coefficient is given by:

$$Cn = \frac{N!}{n!(N-n)!} \qquad (51)$$

Equating the coefficients Cn and the reflection coefficients leads to:

$$\Gamma n = ACn \qquad (52)$$

It can be shown that the amplitude factor is given by:

$$A = 2^{-N}\frac{ZL - Zo}{ZL + Zo} \qquad (53)$$

Furthermore, the adjacent sections of the binomial transformer are related by:

$$\Gamma n = ACn = \frac{Z_{n+1} - Z_n}{Z_{n+1} + Z_n} \qquad (54)$$

or,

$$Z_{n+1} = Z_n \frac{1 + ACn}{1 - ACn} \qquad (55)$$

It can also be shown that:

$$Z_{n+1} = Zo * \exp(K) \qquad (56)$$

where K is:

$$\sum C_k 2^{-N} \ln(\frac{ZL}{Zo}) \qquad (57)$$

The summation is from k = 0 to n.

To further clarify these concepts an example is presented below.

Practical Impedance Matching Techniques

Let us assume we want to design a 3 – section binomial matching transformer to match a 100 ohm load to a 50 ohm line and find the percentage bandwidth for an overall reflection coefficient of 0.3

Solution:

Step 1.0

Find the binomial coefficients from the identity $Cn = \dfrac{N!}{n!(N-n)!}$.

Thus:

$$C0 = \frac{3.2.1}{1.3.2.1} = 1$$

$$C1 = \frac{3.2.1}{1.2.1} = 3$$

$$C2 = \frac{3.2.1}{2.1.1} = 3$$

$$C3 = \frac{3.2.1}{3.2.1.1} = 1$$

Note: 0! (factorial 0) is actually equal to 1.0

Step 2.0   Calculate $\ln[ZL/Zo] = \ln[(100/50] = \ln 2 = 0.6931$.

Step 3.0   Calculate equation   (57)

$$\frac{\sum C_k * 0.6931}{8.0} = \sum C_k * 0.0866 \qquad (58)$$

or,

Practical Impedance Matching Techniques

$Z_{n+1} = 50*0.0866*\sum C_k$ (summation is from k=0 to n)

Therefore,

$Z1 = 4.330*1 = 4.330$ Ohms , here $\sum C_k = C0$ since n = 0
$Z2 = 4.330 * 4 = 17.32$ Ohms , here $\sum C_k = C0 + C1$ since n = 1
$Z3 = 4.330 * 7 = 30.31$ Ohms , here $\sum C_k = C0 + C1 + C2$ since n = 2

Note that all lengths are quarter wave long.

4.4.2 The percentage bandwidth for the binomial transformer:

The percentage bandwidth is given by the following expression:

$$\frac{\Delta f}{fo} = 2 - (\frac{4}{\pi})\cos^{-1}[0.5*(\frac{\Gamma_m}{|A|})^{\frac{1}{N}}]$$

(59)

and ,

$$A = 2^{-N} \frac{ZL - Zo}{ZL + Zo}$$

(59.1)

Γ_m = overall reflection coefficient at its maximum allowable value, *a specified value between 0 and 1 obviously.*

Again using the expression (59) and (59.1) above, the percentage bandwidth can be calculated. The example shows the value for the parameters chosen there.

The percentage bandwidth for the example above is:

$$A = \frac{1}{8}(\frac{50}{150}) = 0.04 = \frac{1}{24}$$

$$Bw\% = 2.0 - (1.27)\cos^{-1}[0.5*(\frac{2.4}{1})^{0.33}] = 0.72 \ (72\%)$$

4.4.3 Multisection matching with a Chebyshev transformer:

A still broader bandwidth can be achieved with a multisection circuit using impedances based on Chebyshev functions. The trade-off is the ripple in the passband. However, even in this case we can still specify a certain maximum allowable reflection coefficient. The following describes the design.

Chebyshev polynomials have the following properties:

1.0 Even ordered Chebyshev polynomials are even functions.
2.0 Odd ordered Chebyshev polynomials are odd functions.
3.0 The magnitude of any Chebyshev polynomial is unity or less than unity in the range of $-1 \leq x \leq 1$ where x is the independent variable of the polynomial.
4.0 $Tn(1) = 1$ for all Chebyshev polynomials.
5.0 All the roots of Chebyshev polynomials lie in the range of $-1 \leq x \leq 1$.

Chebyshev polynomials up to the ninth order are listed below

$T0(x) = 1$
$T1(x) = x$
$T2(x) = 2x^2 - 1$
$T3(x) = 4x^3 - 3x$
$T4(x) = 8x^4 - 8x^2 + 1$
$T5(x) = 16x^5 - 20x^3 + 5x$
$T6(x) = 32x^6 - 48x^4 + 18x^2 - 1$
$T7(x) = 64x^7 - 112x^5 + 56x^3 - 7x$
$T8(x) = 128x^8 - 256x^6 + 160x^4 - 32x^2 + 1$

$T9(x) = 256x^9 - 576x^7 + 432x^5 - 120x^3 + 9x$

In order to facilitate design it is useful to write the Chebyshev polynomials in terms of $x = \cos\theta$. Once this is done, the transformed polynomials may be written:

$T0(\theta) = \cos(0) = 1.0$
$T1(\theta) = \cos(\theta) = \cos(\theta)$
$T2(\theta) = \cos(2\theta) = 2\cos^2(\theta) - 1$
$T3(\theta) = \cos(3\theta) = 4\cos^3(\theta) - 3\cos(\theta)$
$T4(\theta) = \cos(4\theta) = 8\cos^4(\theta) - 8\cos^2(\theta) + 1$
$T5(\theta) = \cos(5\theta) = 16\cos^5(\theta) - 20\cos^3(\theta) + 5\cos(\theta)$
$T6(\theta) = \cos(6\theta) = 32\cos^6(\theta) - 48\cos^4(\theta) + 18\cos^2(\theta) - 1$
$T7(\theta) = \cos(7\theta) = 64\cos^7(\theta) - 112\cos^5(\theta) + 56\cos^3(\theta) - 7\cos(\theta)$
$T8(\theta) = \cos(8\theta) = 128\cos^8(\theta) - 256\cos^6(\theta) + 160\cos^4(\theta) - 32\cos^2(\theta) + 1$
$T9(\theta) = \cos(9\theta) = 256\cos^9(\theta) - 576\cos^7(\theta) + 432\cos^5(\theta) - 120\cos^3(\theta) + 9\cos(\theta)$

or a compact notation is:

$$\Gamma n(\cos\theta) = \cos n\theta \qquad (60)$$

Also Chebyshev polynomials for all arguments may be written as,

$Tn(x) = \cos(n\cos^{-1}x)$ for $|x| < 1$ (61)
$Tn(x) = \cosh(n\cosh^{-1}x)$ for $|x| > 1$ (62)

Practical Impedance Matching Techniques

In order to use Chebyshev polynomials, the end points of the passband given by $(\theta_m, \pi - \theta_m)$ with the center frequency at $\pi/2$, must be mapped onto the range where the Chebyshev polynomials satisfy the identity,

$$|T_n(\cos\theta)| \leq 1.$$

Such a mapping exists and is given by:

$$T_n\left(\frac{\cos\theta}{\cos\theta_m}\right) = T_n(\sec\theta_m \cos\theta)$$

Using this mapping the polynomials can be written as:

$$T_0(\sec\theta_m \cos\theta) = 1$$
$$T_1(\sec\theta_m \cos\theta) = \sec\theta_m \cos\theta,$$

and so on.

The reflection coefficient of a N multisection transformer is given by:

$$\Gamma(\theta) = 2\, e^{-jN\theta}\left[\Gamma_0 \cos N\theta + \Gamma_1 \cos(N-2)\theta + \ldots + \Gamma_n \cos(N-2n)\theta + \ldots\right] \tag{62.1}$$

which reduces to,

$$\Gamma(\theta) = A e^{-jN\theta} T_N(\sec\theta_m \cos\theta) \tag{63}$$

The maximum magnitude of the reflection coefficient in the passband is A.

This is because the maximum magnitude of the Chebyshev polynomial cannot exceed unity in the passband.

Practical Impedance Matching Techniques

The value of A is given by:

$$A = \frac{\ln(\frac{ZL}{Zo})}{2\Gamma_N (\sec\theta_m)} \qquad (64)$$

(This result is derived by taking the limit of the reflection coefficient $T(\theta)$ as θ approaches zero. Also because $\Gamma(0) = \frac{ZL - Zo}{ZL + Zo}$

Using this equation we can also write for the reflection coefficient:

$$\Gamma(\theta) = \frac{\ln(\frac{ZL}{Zo})}{2\Gamma_N (\sec\theta_m)} \cdot e^{-jN\theta} \Gamma_N(\sec\theta_m \cos\theta) \qquad (65)$$

from equations (63) and (64).

The angle θ_m can be found from the expression below ((presented without proof).

$$\sec(\theta_m) = \cosh\left[\frac{1}{N}\cosh^{-1}\left|\frac{\ln(ZL/Zo)}{2\Gamma_m}\right|\right] \qquad (66)$$

The characteristic impedances can be found from,

$$Z_{n+1} = Z_n e^{2\Gamma_m} \qquad (67)$$

where the Γ_m are the reflection coefficients.

The fractional bandwidth is,

$$\frac{\Delta f}{f_o} = 2 - \frac{4\theta_m}{\pi} \qquad (68)$$

The following example is presented to clarify these design concepts.

A 4 section Chebyshev transformer is required to match a 300 Ohm load to a 50 Ohm line. The maximum reflection coefficient cannot exceed 0.1.

From eqn (66) we get,

$$\sec\theta_m = \cosh\left[\frac{1}{4}\cosh^{-1}\left|\frac{\ln(6)}{2(0.1)}\right|\right] \quad \text{which gives } \theta_m = 38.2 \text{ Degrees.}$$
$$(69)$$

The reflection coefficient for a fourth order section is given by:

$$\Gamma(\theta) = A e^{-j4\theta}\, T_4(\sec\theta_m \cos\theta) \qquad (70)$$

Now we can replace T_4 by its expanded version:

$$T_4(\sec\theta_m \cos\theta) =$$
$$\sec^4\theta_m(\cos 4\theta + 4\cos 2\theta + 3) - 4\sec^2\theta_m(\cos 2\theta + 1) + 1$$
$$(71)$$

Also the equation for the reflection coefficient with $N = 4$ is:

$$\Gamma(\theta) = 2\, e^{-j4\theta}\left[\Gamma_0 \cos 4\theta + \Gamma_1 \cos 2\theta + .\Gamma_2/2\right] \qquad (72)$$

Now if equation (71) is substituted in equation (70) and after some manipulation the coefficients of the cosine terms are equated, we get:

$$\Gamma_o = 0.131,$$
$$\Gamma_1 = 0.201$$
$$\Gamma_2 = 0.239$$

Now additionally, because of symmetry:

$$\Gamma_4 = \Gamma_0$$
$$\Gamma_3 = \Gamma_1$$

Therefore, using equation (67) we get;

$$Z_1 = Z_0 e^{2\Gamma_0} = 65 \text{ Ohm}$$

Similarly,

$$Z_2 = 97.1 \quad \text{Ohm}$$
$$Z_3 = 156.7 \quad \text{Ohm}$$
$$Z_4 = 234.2 \quad \text{Ohm}$$

Transmission line expressions and formulas

4.5.1 Transmission line facts:

Before going any further in this discussion the following facts about transmission lines should be considered:

0.0 Transmission lines that have electrical lengths *less than 90 Degrees* behave *inductively for short circuited loads* and *capacitively for open circuited loads.*

1.0 A *RF short circuit* can be produced *at any point* in a circuit by using a *short circuited transmission line with a half wavelength electrical length*. This short repeats itself every multiple of a half wavelengh.

2.0 A transmission line's input impedance is always equal to the termination at adjacent ends of the line if the characteristic impedance of the line is *the same as* the termination.

3.0 Transmission lines have *large transformation capabilities*. A short circuit can be transformed to an open circuit by using a *90 Degree* long line. The same is true of an open circuit. An open circuit can be transformed into a short circuit by a *90 Degree* line.

4.0 Cascaded transmission lines form *concentric circles* on a normalized Smith Chart if the load impedance is normalized to the characteristic impedance of the transmission line.

5.0 Parallel open and short circuited stubs behave *inductively or capacitively* as long as their electrical length is *less than 90 Degrees*.

6.0 Parallel stubs *always* move on the constant conductance circles on the Smith Chart.

7.0 A correctly selected combination of a *parallel stub and cascade transmission line* can be used to *transform any point* on the Smith Chart to any other point. The topology of this combination is dependent on the relationship of the two points. Sometimes the cascade line can be used first followed by the stub, while at other times the stub is followed by the cascaded line.

4.5.2 Electrical length

Sooner or later, the design engineer who is working in microwave or high frequency electronics, is going to come up against the concept of electrical length. In order to understand this concept lets work out the following arithmetic:

1.0 The wave number* or phase constant = $\beta = 2\pi/\lambda$

2.0 The electrical length is defined by $\theta = \beta l$ where l = physical length.

3.0 $\theta = \beta l = (l/\lambda) *360$ degrees

Here λ is the wavelength of the signal in the applicable dielectric (or sometimes called the guide wavelength).

4.0 For frequencies in Ghz, this becomes: [360 * fGhz * l(cm) * $\sqrt{\varepsilon eff}$]/30 cm

In this case frequency is in Ghz, physical length is in centimeters.

For example:

Let frequency be 1 Ghz.
Let $\lambda = 0.8\ \lambda$(air) or $\sqrt{\varepsilon eff} = 1.25$
Let $l = 0.1$ meters = 0.1E2 centimeters.
Then :
$\theta = [360* 1*0.1E2*1.25]/30$ degrees

- See the Signal Processing Group Inc. blog located at:
  http://signalpro.biz/wordpress

Transmission line parameters and characterization

In this section we present some parameters of transmission lines useful to the practicing engineer or student of transmission line matching

The Propagation Constant:

A traveling wave on a transmission line has a voltage v and current i. These two quantities are related by the *Characteristic Impedance* of the line as:

$$Zo = \frac{v}{i} = \sqrt{\frac{Ro + j\omega Lo}{Go + j\omega Co}} \qquad (S24.0)$$

where:

Ro = resistance per unit length of the line.

Go = shunt conductance per unit length of the line.

Lo = Series inductance per unit length of the line.

Co = shunt capacitance per unit length of the line.

The equivalent RLCG circuit of a transmission line is shown below.

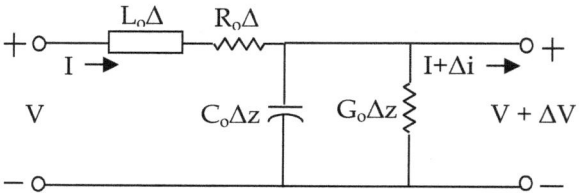

Practical Impedance Matching Techniques

The propagation constant of a real, lossy transmission line is given by:

$$\gamma = \alpha + j\beta \qquad (S25.0)$$

which is also given by:

$$\sqrt{(Ro+j\omega Lo)(Go+j\omega Co)} \qquad (S26.0)$$

Here,

α = attenuation constant in nepers/unit length

β = phase constant in radians/unit length

For the ideal line the propagation constant may also be written:

$$\gamma = j\omega\sqrt{(LoCo)} \qquad (S26.1)$$

For <u>low loss lines,</u>

$$\alpha = \frac{Ro}{2Zo} + \frac{GoZo}{2} \qquad (S26.2)$$

$$\beta =$$

$$\omega\sqrt{LoCo}\left\{1.0 - \left[\frac{RoGo}{4\omega^2 LoCo}\right] + \left[\frac{Go^2}{8\omega^2 Co^2}\right] + \left[\frac{Ro^2}{8\omega^2 Lo^2}\right]\right\}$$

(S26.3)

For <u>the ideal line,</u>

$$\alpha = 0.$$

$$\beta = \omega\sqrt{L_o C_o} = \frac{\omega}{v} = \frac{2\pi}{\lambda} \qquad (S26.4)$$

Phase velocity:

The phase velocity is given by:

$$v = \frac{\omega}{\beta} \qquad (S27.0)$$

Another parameter used commonly is the velocity factor given by:

$$VF = 1/(\sqrt{\mu_r \varepsilon_r})$$

where,

μ_r = relative permeability, ε_r = relative permittivity

Characteristic impedance:

If the lossy components of the line are essentially zero, i.e. the line is considered lossless, then the line characteristic impedance can be defined as:

$$Z_o = \sqrt{\frac{L_o}{C_o}} \qquad (S28.0)$$

or generally,

$$Z_o = \sqrt{\frac{R_o + j\omega L_o}{G_o + j\omega C_o}} \qquad (S28.1)$$

For low – loss lines,

$$Z_o = \sqrt{\frac{L_o}{C_o}}\left\{1.0 + j\left[\left(\frac{G_o}{2\omega C_o}\right) - \left(\frac{R_o}{2\omega L_o}\right)\right]\right\} \quad (S28.2)$$

Electrical length:

This term refers to the ratio of the physical length '*l*' of the transmission line to the wavelength, λ in the applicable dielectric. The wavelength in the applicable dielectric, sometimes called the guide wavelength is given by:

$$\lambda_G = \frac{\lambda_o}{\sqrt{\varepsilon_{eff}}} \quad (S29.0)$$

Here:

ε_{eff} = Effective dielectric constant , and λ_o = wavelength in air

If the signal wave propagates in homogeneous media then the effective dielectric constant is equal to ε_R, the relative dielectric constant. However, in cases of microstrip line, for instance, where part of the wave is in the air and part in the dielectric, the effective dielectric constant is *not equal* to the relative dielectric constant. In such a case the effective dielectric constant has to be calculated.

Fractional wavelength:

Fractional wavelength is the ratio of the physical length of the line to the effective or guide wavelength.

Fractional wavelength (%) $= \left[\dfrac{l}{\lambda_G}(100) \right]$ (S30.0)

or,

In degrees:

θ(Degrees) $= [l/\lambda_G] [360]$ (S31.0)

Please note that in the above equations, the quantity l can be confused with the numeral 1. Thus l/λ_G is length/guide wavelength, not one divided by guide wavelength.

Input impedance:

An ideal transmission line terminated with a load of ZL has an input impedance of:

$$ZIN = Zo \dfrac{ZL + jZo \tan(\theta)}{Zo + jZL \tan(\theta)} \quad (S32.0)$$

For a general transmission line the input impedance expression is:

$$ZIN = Zo \dfrac{ZL + jZo \tanh(\gamma l)}{Zo + jZL \tanh(\gamma l)} \quad (S33.0)$$

where ZL = the load or termination and Zo is the characteristic impedance.

Impedance of a shorted line:

This is also addressed in the section on stubs. In any case, the impedance of a *ideal shorted line* is:

($\theta = \beta l$, the electrical length)

$$Z = jZo\tan(\theta) \qquad (S34.0)$$

while for *a general line:*

$$Z = Zo\tanh(\gamma l) \qquad (S35.0)$$

and for *a low-loss line:*

$$Z = Zo\frac{\alpha l + j\tan\theta}{1.0 + j\alpha l\tan\theta} \qquad (S36.0)$$

Impedance of an open line:

For an *ideal line:*

$$Z = -jZo\cot\theta \qquad (S37.0)$$

For a *general line,*

$$Z = Zo\coth(\gamma l) \qquad (S38.0)$$

For a *low – loss line:*

$$Z = Zo\frac{1 + j\alpha l\tan\theta}{\alpha l + j\tan\theta} \qquad (S39.0)$$

Impedance of a quarter-wave line:

For an *ideal line:*

$$Z = \frac{Zo^2}{ZL} \qquad (S40.0)$$

For a *general line,*

$$Z = Zo\frac{ZL + Zo\coth \alpha l}{Zo + ZL\coth \alpha l} \qquad (S41.0)$$

For a *low – loss line:*

$$Z = Zo\frac{Zo + ZL\alpha l}{ZL + Zo\alpha l} \qquad (S42.0)$$

Impedance of a half-wave line:

For *an ideal line:*

$$Z = ZL \qquad (S43.0)$$

For *a general line:*

$$Z = Zo\frac{ZL + Zo\tanh \alpha l}{Zo + ZL\tanh \alpha l} \qquad (S44.0)$$

For a *low-loss line:*

$$Z = Zo\frac{ZL + Zo\alpha l}{Zo + ZL\alpha l} \qquad (S45.0)$$

Voltage along the line:

For the *ideal line:*

$$V_{line} = V_{input}\cos(\beta z) - jI_{input}Z_o\sin(\beta z) \quad (S46.0)$$

where:

Vinput is the input voltage and,

Iinput is the input current.

For a *general line*:

$$V_{line} = V_{input}\cosh(\gamma z) - I_{input}Z_o\sinh(\gamma z) \quad (S47)$$

Current along the line:

For the ideal line: $I_{line} = I_{input}\cos(\beta z) - j\dfrac{V_{input}}{Z_o}\sin(\beta z)$

(S48.0)

For the general line:

$$I_{line} = I_{input}\cosh(\gamma z) - \dfrac{V_{input}}{Z_o}\sinh(\gamma z) \quad (S49.0)$$

The reflection coefficient:

For both an *ideal line* and a *general line*:

$$\Gamma = \dfrac{Z_L - Z_o}{Z_L + Z_o} \quad (S50.0)$$

The standing wave ratio:

For both the ideal line and a general line,

SWR = [1+ $|\Gamma|$]/[[1-$|\Gamma|$]]   (S51.0)

Microstrip lines.

Microstrip lines are used commonly both in IC design, and in PCB design. Some interesting features of these types of transmission lines are described below. For a really helpful script use *Appcad scripts*

A cross section of the microstrip line is shown below. As shown, it consists of a top conducting metallic strip, an insulating substrate and a ground plane on the bottom of the substrate.

Once the Zo of the line is known most of the above formulas can be used in synthesis and analysis of the impedance matching networks

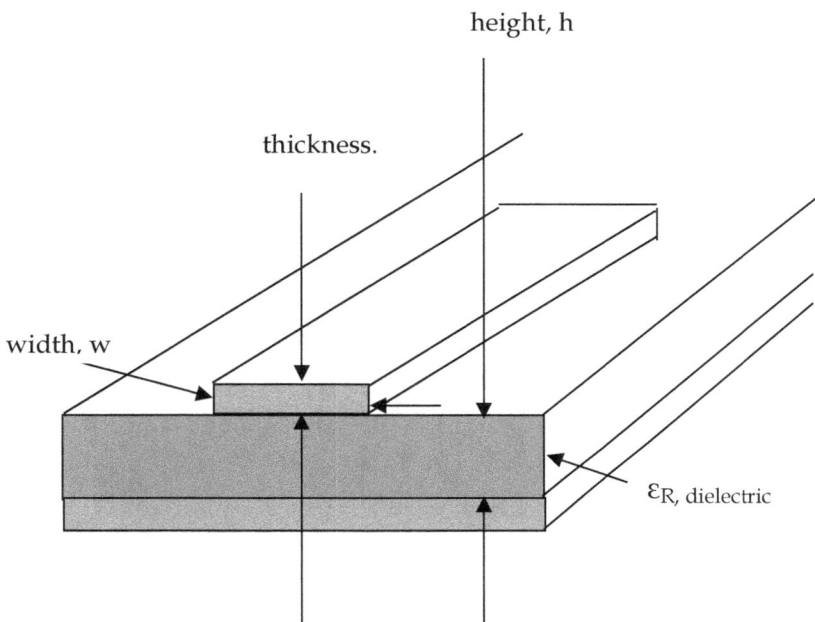

Series stubs or short high impedance (narrow) lengths of line are inductive and wide short lengths of microstrip are capacitors.

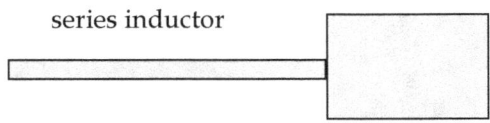

In addition a gap in the microstrip line can be assumed to be a series capacitor. This is shown below.

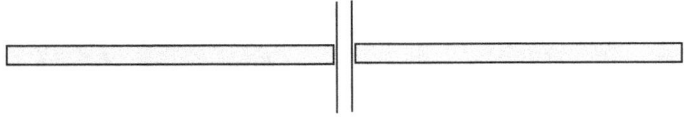

Series gap, series capacitor

Other type of components using microstrip lines have been described above.

Microstrip design for matching elements and stubs.

To begin with, note that microstrip structures are such that the electromagnetic fields are partly in air and partly in the dielectric medium it is made of. Thus microstrip's EM mode of propagation is called a quasi TEM mode. The effective permittivity of microstrip is affected significantly. A number of expressions have been generated to calculate it. In this discussion we will use the following expressions for effective permittivity. (Please note if you have downloaded APPCAD you can use it to do many of these permittivity and Zo calculations quickly.)

$$\varepsilon eff = [(\varepsilon r+1)/2][(\varepsilon r-1)/2][1.0+10(h/w)]^{-0.555} \qquad M.1.0$$

where w/h > 1.3 Zo>(63-2εr) Ohms.

In this equation:

> h = thickness of the substrate
> w = width of the microstrip
> Zo = characteristic impedance of the microstrip
>
> This equation is accurate to within 1%.

Note that as the width of the microstrip becomes wide compared to its thickness the microstrip starts to approach an ideal parallel plate capacitor with $\varepsilon\text{eff} = \varepsilon r$. εr is the relative permittivity of the substrate. For example FR-4 has a relative permittivity of approximately 4.6.

Another popular expression for effective permittivity is:

$$\varepsilon\text{eff} = [(\varepsilon r+1)/2]+[(\varepsilon r-1)/2][1.0/\sqrt{\{1.0+(12h/w)\}}]$$
$$\text{M2.0}$$

Note that these expressions are valid for frequencies in the lower microwave region. (Few gigahertz and below) If calculations are required for higher microwave frequencies please download the freeware to generate the higher frequency expressions from the Signal Processing Group Inc. website (http://www.signalpro.biz.) At higher frequencies a number of side effects have been noted such as dispersion.)

However, these expressions can be used to get a starting point for a design and then CAD tools (if available) can be used to fine tune the circuits. In the absence of CAD tools the reader may access calculators from Signal Processing Group Inc. website for more accurate calculations of the various quantities involved.

Calculators for accurate Zo values may be found in the accompanying software of the book. There are many other useful calculators also available. Please see the end of the book for ways to access these calculators.

Zo also is affected by higher frequencies and its effects are embodied in freeware available from the Signal Processing Group Inc website under complementary items.

The combination of this book, the freeware and calculators can be very useful to designers when they are starting a design or optimizing the design.
Additional free calculators are available from APPCAD (Avago freeware) and TXLINE (from National instruments, also freeware). These, too, are very useful and should be used to verify hand calculations.

Another quantity that has been mentioned elsewhere in this book (and the first edition) is the guide wavelength or effective wavelength that is simply given by:

$\lambda g = \lambda / \sqrt{(\varepsilon eff)}$ M.3.0

We will now address the various elements that are useful in matching and passive microstrip circuits.

1.0   Inductor: An inductor is formed when a narrow microstrip line is used. (An example is a microstrip line of 100 Ohms characteristic impedance). In this case the equivalent circuit of this line is a pi circuit with capacitors at the two ends of the inductor and an inductor. See figure M.1 below.

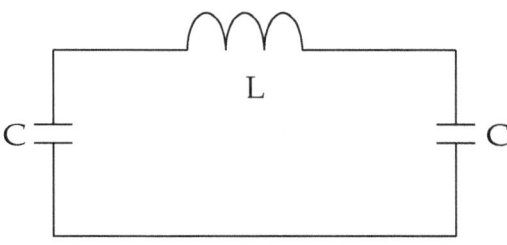

Figure M.1

The inductor value may be calculated from the following expression.

XL = inductive reactance = Zo.Sin(EL).   M4.0

Here EL = Electrical length = 2π.length/λg.
length is the physical length of the strip.

This expression allows us to calculate the physical length required for a strip to generate a required XL. The width is given by the Zo of the line.

The capacitors at the two ends may be calculated using the following expression.

BL = (1/Zo)Tan(π.length/λg.)   M.5.0

See figure M.2 below for a conceptual illustration of the structure.

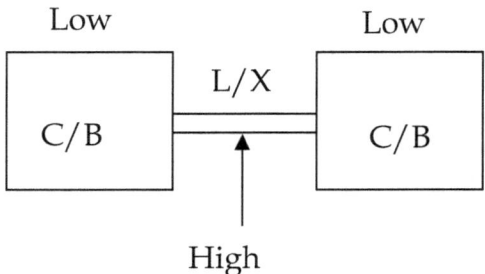

Figure M.2

2.0 Capacitor: In a similar fashion a capacitor can be realised using a section of mircostrip line. In this case the line is made wide and of large area. This structure is shown below in Figure M.3.

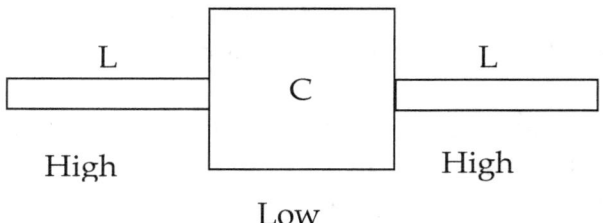

Figure M.3

In this case the equivalent circuit looks like Figure M.4

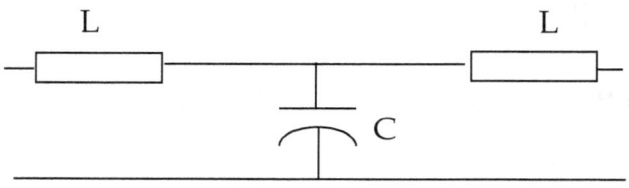

Figure M.4

The capacitance can be calculated from:

B = shunt susceptance = $(1/Z_o)\mathrm{Sin}(EL)$.   M.6.0

The parasitic inductance at the two ends can be calculated from :

XL_{para} = series reactance = $(\pi \cdot length \cdot Z_o)/\lambda_g$ M.7.0

This parasitic inductance is very small. However, even though the reactance is small these inductive lines add to the capacitance of the center section. So, in fact the total capacitance of this structure is the center section capacitance plus the capacitance of the feeder lines which also provide capacitance as well as the steps in width of the lines connecting to them. So these corrections should be made. The expressions for these have already been provided above.

To summarize these findings we present the tables below for both the inductive calculations and the capacitive calculations and provide the calculators to find these values quickly. Please refer to the calculators available with the book. These can be acquired separately as needed.

Table M.1

Inductive	Capacitive
$X_L = Z_o \cdot \sin(EL)$	$B = (1/Z_o) \cdot \sin(EL)$
End capacitance: $B_L = (1/Z_o)\tan(\pi \text{length}/\lambda_g)$	Feeder inductances: $X_{Lpara} = \pi \cdot \text{length} \cdot Z_o/\lambda_g$
Physical length of the inductive line: $\text{length} = (\lambda_g/2.0.\pi)\sin^{-1}(\omega L/Z_o)$	Physical Length of capacitive section: $\text{length} = (\lambda_g/2.0.\pi)\sin^{-1}(\omega C Z_o)$

Open and shorted microstrip line characteristics: Let us examine how open and shorted microstrip lines behave based on their lengths as a function of effective wavelength of the signal.

Open line:
When a line is open ended then at all *odd* quarter – wave points (1/4, 3/4 etc) the impedance is *minimum*. The line acts like a series resonant circuit. So if one wishes to establish a RF short circuit to ground (for example) at some point on a microstrip line, then an open ended, quarter-wave (or odd multiples) of an open ended line connected to that point will drive the line to very low impedance and short the frequency that is used to ground. See Figure M.5 below.

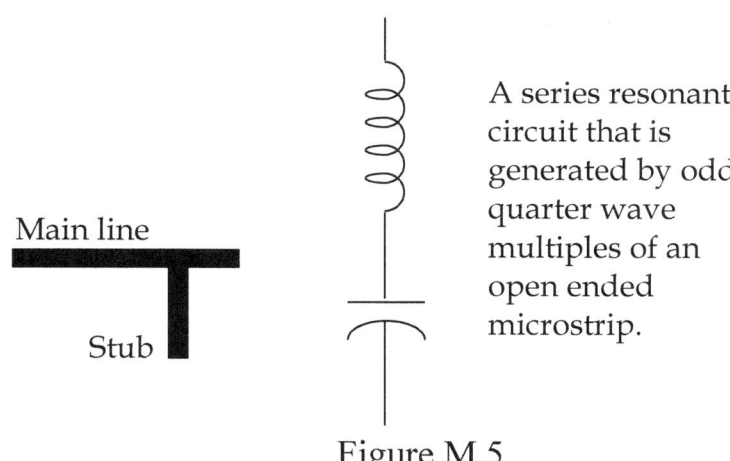

Figure M.5

At all *even* quarter – wave points (1/2, 1, 3/2 etc) the impedance is a *maximum.* The line behaves as a parallel resonant circuit. The line at resonance has a very high impedance. Please refer to Figure M.6 below.

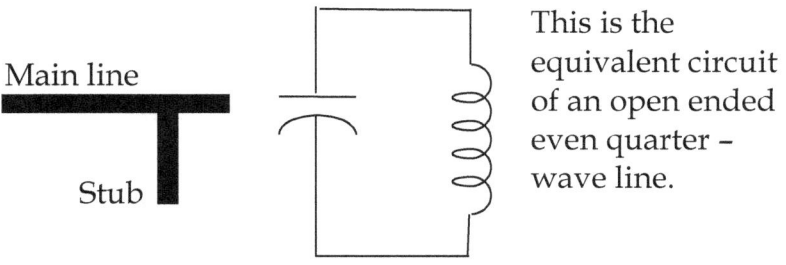

Figure M.6

An earlier discussion in this book addressed open and shorted stubs that are less than a quarter wavelength long. An open ended line that is less than a quarter – wave long acts as a capacitance.

A open ended line that is between a 1/4 to a 1/2 wavelength long acts as an inductance. An open ended line that is 1/2 to 3/4 wavelength long acts as a capacitance etc.

Shorted line:

When the line is shorted at one end (RF short) then at *odd* quarter – wavelength points, it acts like a parallel resonant tuned circuit. Thus at the frequency of resonance the impedance is very high.

At *even* quarter - wave points the line acts as a series resonant circuit and generates a low impedance at the resonant frequency

Resonant shorted lines also can act as pure capacitances and inductances. A shorted line that is less than a 1/4 wavelength long acts as an inductance. A shorted line that is between a 1/4 to a 1/2 wavelength long acts as a capacitance. From 1/2 to 3/4 wavelength it acts as an inductance and from 3/4 to 1 wavelength long it acts as a capacitance and so on.

It should be clear that appropriately chosen line lengths can be used as parallel – resonant or series – resonant , inductive or capacitive circuits.
This is a very powerful conclusion and is very useful for the design engineer.

Open circuit microstrip lines and terminations: The following are relevant facts:

O1.0 The voltage at the open end is maximum but the current is minimum.
O2.0 The distance between two adjacent zero current points is 1/2 wavelength.

Practical Impedance Matching Techniques

O3.0   Distance between two alternate zero current points is 1 wavelength.

O4.0   At any frequency the voltage on the line is minimum or zero at 1/4 wavelength from the end of the line.

O5.0   Voltage peaks occur at the end of the line, at 1/2 wavelength from the end and every 1/2 wavelength after that.

Short circuit microstrip lines and terminations: The following are relevant facts:

S1.0   The voltage at the end (terminated end) is obviously zero.

S2.0   The voltage is maximum at 1/4 wavelength from the end and alternately maximum at every 1/4 wavelength after that.

Microstrip lines terminated in Zo: The following are relevant facts:

Z1.0   In a properly terminated line the voltage and current will be constant along the line unless there are some losses such as resistive or otherwise.

Z2.0   If there are losses in the line (and there are sure to be some) the voltage and current will become smaller along the line towards the termination.

A point worth noting is that if a line is designed to be a 1/4 wave at frequency fo then it would change its characteristics at a frequency 2fo and 1/2fo etc. This fact must be borne in mind. It could be an advantage or a disadvantage depending on the conditions.

Cross – junction stubs: This is a technique that can be used advantageously in the use of stubs. *When the impedance of the stub is really low, i.e it has a large width then one solution is use two stubs in parallel connected on both sides of the main line.* The impedance of each of the equivalent stubs is twice the impedance of the original stub. See figure M.7.0 below.

Figure M.7.0

At this point it is appropriate to investigate impedance matching using stubs. Both single and double stub matching. Before we do that, let us understand:

The Smith Chart and impedance matching techniques

The Smith Chart is a valuable aid in graphically visualizing the reflection coefficient and Y-Z networks and many other associated quantities. It was invented by Phillip Smith in 1932 and it is still being used today. Its principal operations and techniques are discussed in this part of the eCADbook. Impedance matching techniques using it are also presented.

Practical Impedance Matching Techniques

The reader should use the freeware mentioned in the text and use it to clearly see the relationships on the chart in color. This book may be printed in black and white and some of the graphic/descriptive information may not be too clear on the examples provided.

The Smith Chart is a mapping between the normalized *complex impedance plane* given by z = r+jx and *the complex reflection plane*. The normalized impedance is given by,

$$z = Z/Zo = [R + jX]/Zo \quad (S0.0)$$

The right hand side of a *normalized* complex impedance plane is such that the values of an impedance that has a real impedance $R \geq 0$, will be represented by points there.

On the other hand, the complex reflection coefficient plane may be written in its polar form as:

$$\Gamma = |\Gamma|e^{j\angle\Gamma} = \Gamma_r + j\Gamma_i \quad (S1.0)$$

The magnitude of the reflection coefficient always lies between 0 and 1.0. Its angle is measured with respect to the positive real axis (Γ_r).

Theory:

The reflection coefficient is defined by:

$$\Gamma = [ZL - Zo]/[ZL+Zo] \quad (S2.0)$$

Here ZL is the load impedance and Zo is the characteristic impedance of the source or the transmission line.
If we rearrange equation (S2.0),

$$ZL = Zo[1+\Gamma]/[1-\Gamma] \qquad (S3.0)$$

If we divide both sides by Zo we get the mapping stated above.

$$z = r + jx = [1+\Gamma]/[1-\Gamma] \qquad (S4.0)$$

Once we substitute the complex expression for the reflection coefficient and equate the real and imaginary parts we get two equations that represent circles in the complex reflection coefficient plane shown below.

$$[\Gamma_r - r/(1+r)]^2 + [\Gamma_i - 0]^2 = [1/(1+r)]^2 \qquad (S5.0)$$

and,

$$[\Gamma_r - 1]^2 + [\Gamma_i - 1/x]^2 = [1/x]^2 \qquad (S6.0)$$

The first circle is centered at :

$$[r/(1+r), 0] \qquad (S7.0)$$

and its location is *always inside* the unit circle of the complex reflection coefficient plane. The radius of this circle is:

$$1/(1+r) \qquad (S8.0)$$

This circle is always *fully contained* in the unit circle. The radius cannot be greater than unity.

The second circle is centered at:

$$[1, 1/x] \quad (S9.0)$$

and its location is *always outside* the unit circle in the complex reflection coefficient plane.

The centers of both circles will be to the right of the unit circle.

The radius of the second circle is $|1/x|$. This radius can vary from 0 to infinity.

The first circles centered on the real axis represent lines of *constant real part of the load impedance*. To re-iterate, r = constant and x can vary here.

The circles whose centers lie outside the unit circle represent lines of *constant imaginary part* of the load impedance. Here x = constant and r varies.

Circles centered at the match point where $Z_L = Z_o$ or $\Gamma = 0$ are equidistant from the origin (i.e. $|\Gamma|$= constant). These circles are called constant VSWR circles.

Of course,

$$VSWR = Vmax/Vmin = [1+|\Gamma|]/[1-|\Gamma|]$$
$$(S10.0)$$

for reference.

In this equation, Vmax and Vmin are the maximum and minimum amplitudes of the standing waves created by source – load mismatch. (Please see the description of VSWR and standing waves at the beginning of this book)

VSWR can vary from 1 to infinity.

The phase of the reflection coefficient is given by the angle from the right – hand side horizontal axis. The angle of the reflection coefficient can vary from − 180 degrees to + 180 Degrees. Angles above the horizontal axis are *positive* and those

below are *negative*. Please use the Smith Chart program to view its image.

As is well known, once the VSWR is extracted other quantities such as Return Loss can be calculated as well. So VSWR is an important parameter in high frequency design. Please see the definitions of this and other parameters in the first part of this eCADbook.™.

There are some points on the Smith Chart with significant importance. These are listed in the table below and illustrated on the Smith Chart drawing.

Practical Impedance Matching Techniques

Point	Name	Z - plane	Γ - plane	Comments
A	Reference impedance	1 + j0	0.0∠0	Single point
B	Ideal resistance	r+j0	ρ∠0 or ρ∠180	Main diagonal of the chart
C	Ideal capacitive reactances	0 − jx	1.0∠Any -	Lower half of the circumference (No resistance)
D	Ideal inductive reactances	0 +jx	1.0∠Any +	Upper half of the circumference (No resistance)
E	Short circuit	0 +j0	1.0∠180	Single point
F	Open circuit	∞	1.0∠0	Single point
G	Upper half of the chart	r+jx	ρ≤1∠Any+	Inductive half of the chart

H	Lower half of the chart	r – jx	ρ≤1∠Any -	Capacitive half of the chart
I	A specific z value	r+ jx	ρ∠+Angle	Single point
J	Complex conjugate of z value	r - jx	ρ∠- Angle	Single point

TABLE 1.0 Significant points on a Smith Chart

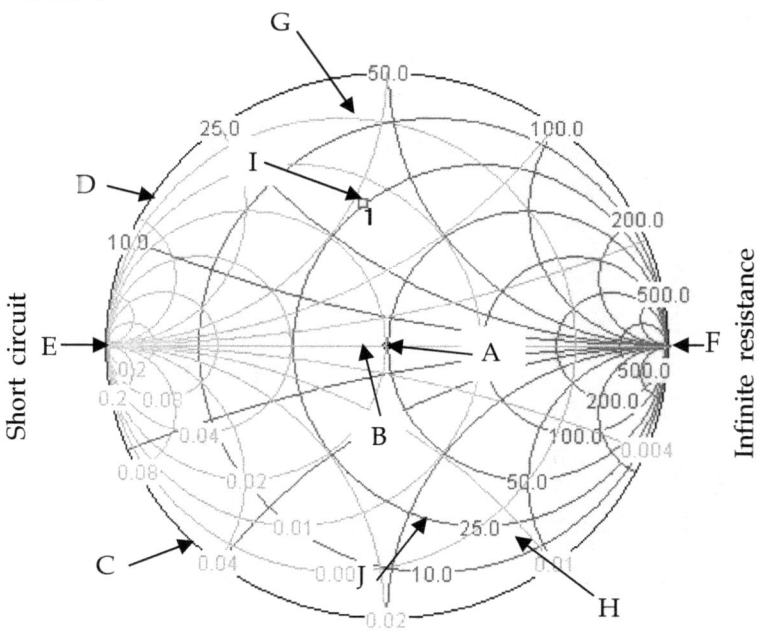

Figure S3.0 Significant points on a Smith Chart:

Impedance manipulations using the Smith Chart.

The Smith Chart can be used in impedance matching. In order to do this, the Smith Chart shows, graphically, the locus of the points when adding series and shunt impedances. This section of the eCADbook™ shows the techniques.

The analysis shown in this section relies heavily on a demo version Smith Chart program. The particulars of the program are:

 V 1.91
This program has been developed by Prof. Fritz Dellsperger, Juerg Tschirren and Roger Wetzel
© 1995 - 2000 by Berne Institute of Engineering and Architecture

It was downloaded from the web as freeware.

- Licence
No valid licence. This copy of 'smith.exe' runs as a DEMOVERSION.

Start with a Smith Chart with a single datapoint. This is where the matching or manipulation starts from. The start is shown below as datapoint "1". This datapoint represents the impedance to be matched to the 50 Ohm (or other impedance) reference load.

Fundamentally datapoint 1, is the impedance to be matched to the reference impedance of 50 Ohms. On the Smith Chart this 50 Ohm impedance lies at the center of the chart as explained and shown above.

The idea is to use "lossless" elements to *move the datapoint to the center of the chart.* Once this is accomplished the matching problem has been solved and the source and the load are conjugate matched to each other.

This is the effort throughout the matching techniques using the Smith Chart.

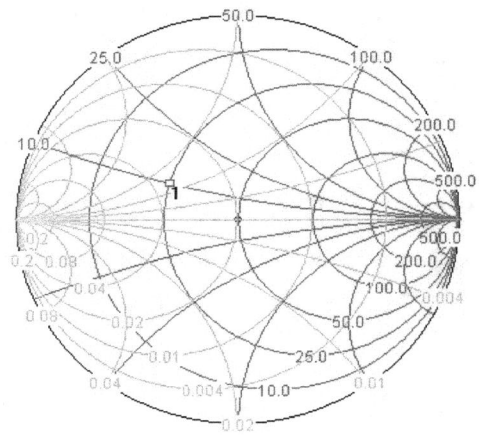

Figure S4.0. Starting from an impedance shown as point "1" on the chart above

This datapoint has the following characteristics: (Reference resistance is 50 Ohm)

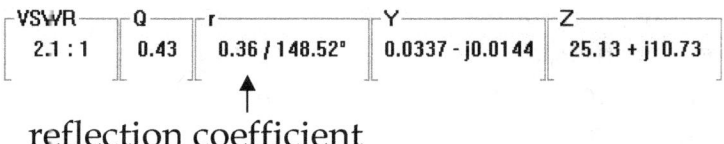

↑
reflection coefficient

We want to add a series inductor to this. This is shown graphically below.

We added a 2.4 nH inductor to the series circuit shown below also. Graphically our datapoint now moves to the new point "2" shown.
Note that the movement is upwards along a circle of positive reactance (i.e an inductor) and a constant resistor.

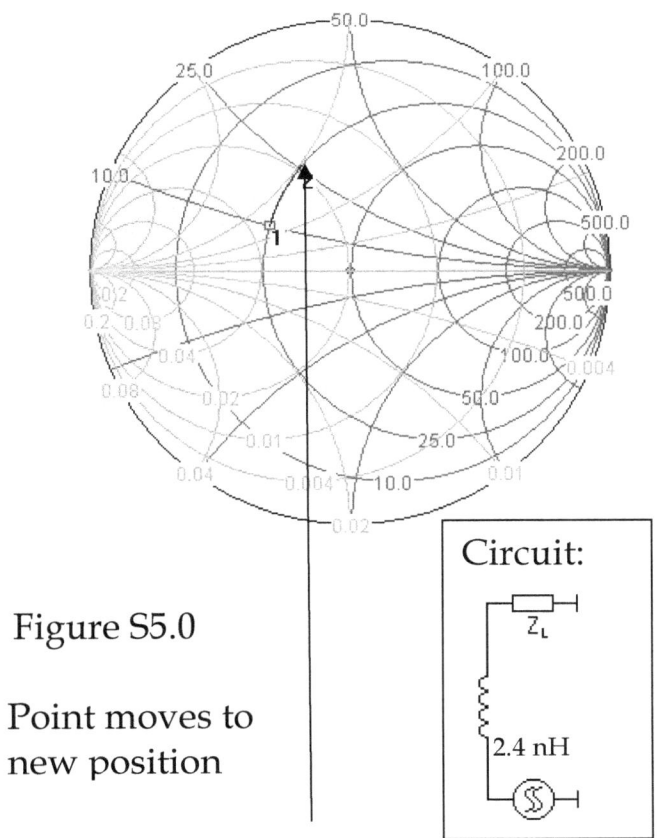

Figure S5.0

Point moves to new position

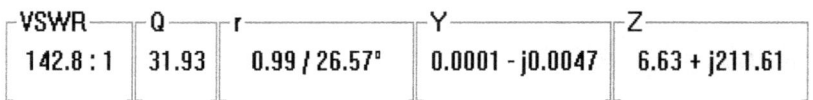

This example shows the addition of a series inductor. If a series capacitor is added as shown below, the trajectory of the datapoint is as shown. Note, the movement is along the negative reactance circle to point "2"

Figure S6.0

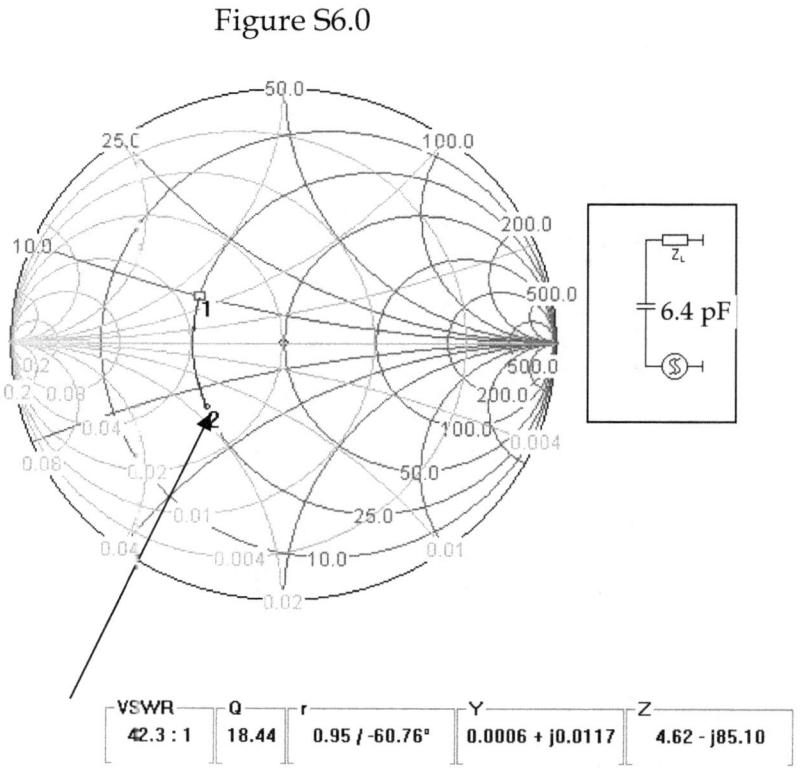

Note the sign of the reactance in the two cases. One is positive (for the addition of an inductor) and the other is negative by virtue of adding a capacitor in series. The resistance is still on a constant circle.

The match is so bad that the VSWR's are in the stratosphere!

Finally a series resistor is added. This moves the datapoint as shown below.

Figure S7.0

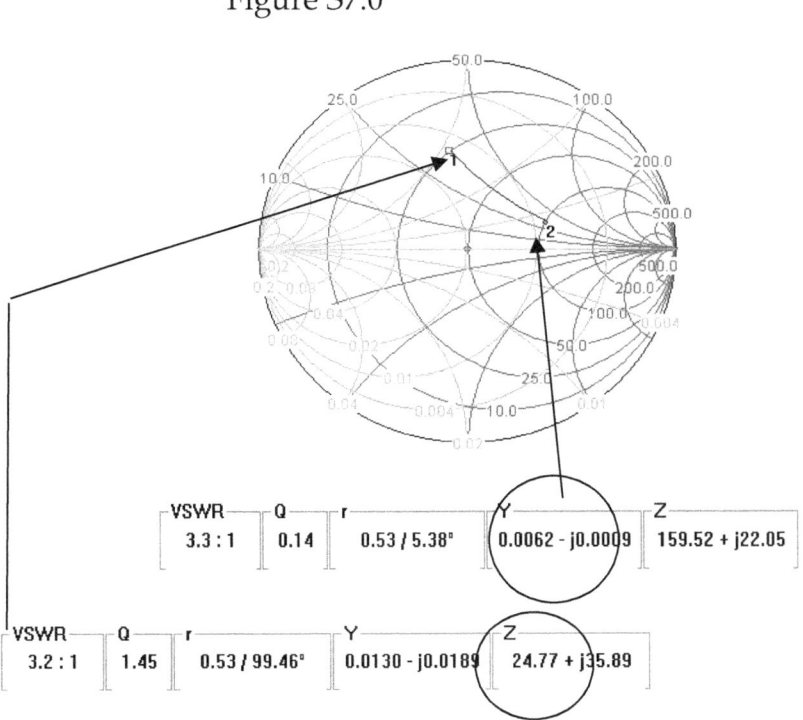

The datapoint has moved from r = 24.77 to r = 159.52 by adding a series resistor.

The movement is from constant resistance circle to constant resistance circle.

Admittance chart:

So far the examples shown have used the impedance chart. The Smith Chart also provides *an admittance chart*. This is shown in the figures of the Smith Chart above.

From the basic theory of normalization:

$\Gamma = [Z - Zo]/Z + Zo] = [1/Y - 1/Yo]/[1/Y + 1/Yo]$ (S11.0)

i.e replace the impedance by the admittance.

Then normalize to get,

$\Gamma = [1 - y]/[1 + y]$ (S12.0)

The normalizing admittance is 1/Rref.

The admittance chart is a 180 degree rotation of the impedance chart by virtue of the expression for impedances above. Here we have $[1 - y]/[1 + y]$ instead of $[z - 1]/[z + 1]$.

Thus the admittance line and circles in the chart below represent the admittance chart. Note, the conductance is the reciprocal of the corresponding impedance.

Practical Impedance Matching Techniques

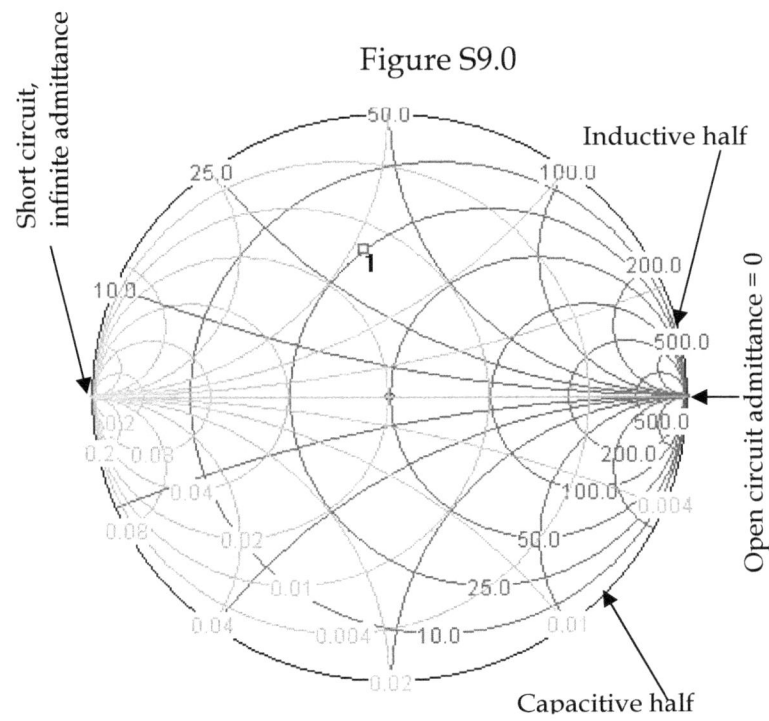

Figure S9.0

Also,

Inductive susceptance is in the top half and capacitive susceptance is in the lower half of the admittance chart. *Inductive susceptances are* negative numbers, while *capacitive susceptances are* positive *numbers.*

The following set of figures shows the effect of adding shunt elements.

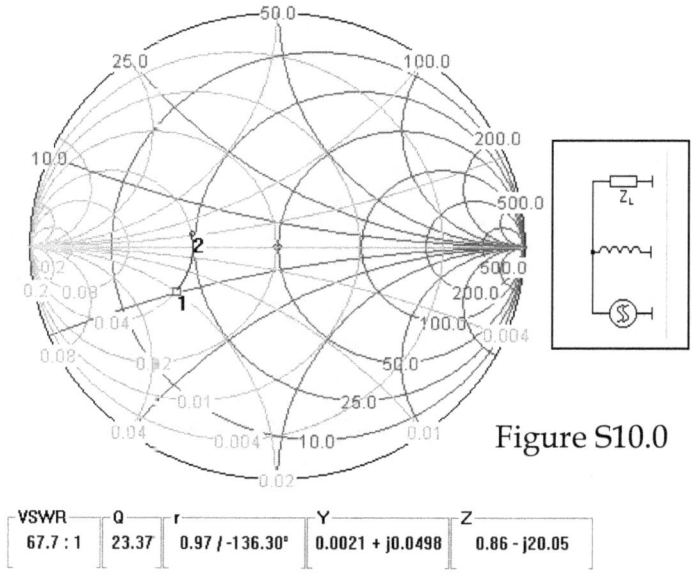

Figure S10.0

VSWR	Q	r	Y	Z
67.7 : 1	23.37	0.97 / -136.30°	0.0021 + j0.0498	0.86 - j20.05

Added shunt inductance. Movement is from "1" to "2". Reference Yo is 0.02

VSWR	Q	r	Y	Z
2.6 : 1	0.50	0.45 / -153.79°	0.0404 + j0.0201	19.84 - j9.89

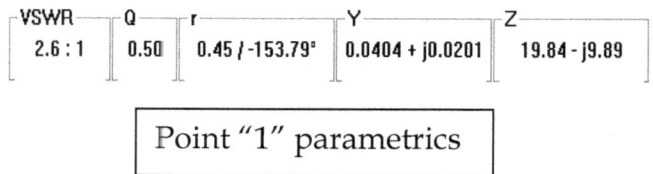

Point "1" parametrics

Practical Impedance Matching Techniques

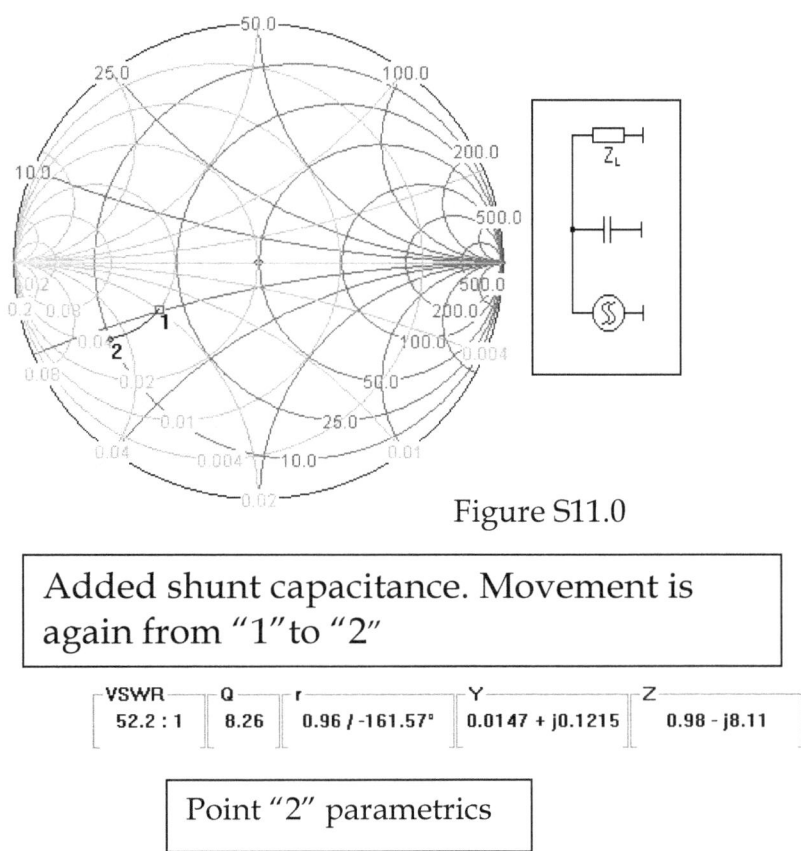

Figure S11.0

Added shunt capacitance. Movement is again from "1" to "2"

VSWR	Q	r	Y	Z
52.2 : 1	8.26	0.96 / -161.57°	0.0147 + j0.1215	0.98 - j8.11

Point "2" parametrics

From the graphics construction above it is clear that *shunt inductors and capacitors move the datapoints on constant conductance circles*

The next graph shows the effect of adding a shunt resistor. Increasing the value of a shunt resistor results in movement of the datapoint towards infinite admittance, i.e a short circuit.

Practical Impedance Matching Techniques

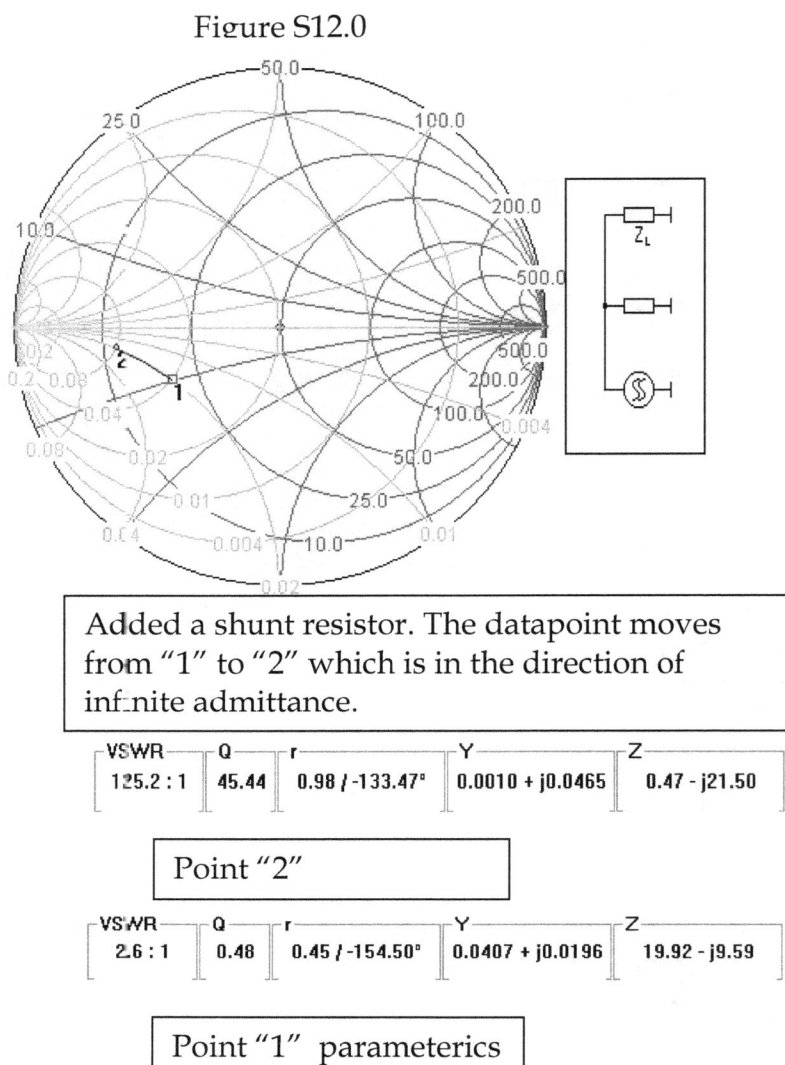

Figure S12.0

Added a shunt resistor. The datapoint moves from "1" to "2" which is in the direction of infinite admittance.

VSWR	Q	r	Y	Z
125.2 : 1	45.44	0.98 / -133.47°	0.0010 + j0.0465	0.47 - j21.50

Point "2"

VSWR	Q	r	Y	Z
2.6 : 1	0.48	0.45 / -154.50°	0.0407 + j0.0196	19.92 - j9.59

Point "1" parameterics

Transmission line manipulations:

One of the interesting features of the Smith Chart, is the manipulations that can be done on transmission lines. Again the following figures show effects of manipulating transmission lines on the Smith Chart.

Cascade transmission line

A few words of explanation are in order at this point when we start examining transmission line manipulations on the Smith Chart. The simplest construct is a *cascade transmission line* shown in the figure below.

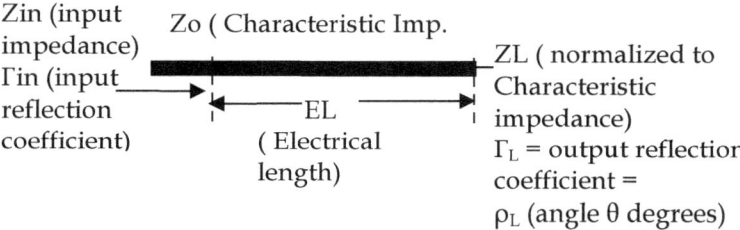

The transmission line has an input impedance given by the familiar equation:

$Z_{IN} = Z_o[Z_L + jZ_o\tan(\theta)]/[Z_o + jZ_L\tan(\theta)]$ (S13.0)

Normalizing this by Zo gives,

$z_{in} = [z_L + j\tan(\theta)]/[1 + jz_L\tan(\theta)]$ (S14.0)

The reflection coefficient is given by (normalized form):

$\Gamma = [z-1]/[z+1]$ (S15.0)

Substituting equation S14.0 into this expression gives, after some manipulation.

$\Gamma_{IN} = \Gamma_L e^{-j2\theta}$ (S16.0)

or,
$$\Gamma_{IN} = \rho_L \angle(\Phi - 2\theta) \qquad (S17.0)$$
Here θ is the electrical length of the line.

The reflection coefficient at the input has the same magnitude as the output reflection coefficient but the phase angle is rotated by Φ in a clockwise direction through 2X the electrical length of the line. This expression applies to any transmission line impedance since the load termination is normalized to the characteristic impedance of the line.

Using the Smith Chart it is possible to find the input reflection coefficient of a terminated line. First we need to normalize the termination with the reference impedance, then mark the load as a datapoint on a normalized Smith Chart. Rotating from the normalized load through an angle that is equal to *2X the electrical length* of the line moves the datapoint to the input reflection coefficient.

The reason the transmission line length is doubled for this method is that a wave applied to the input goes to the output and then comes back to the input again. A distance of 2X the electrical length.

An example is required to see this much more clearly. This is shown below.

Practical Impedance Matching Techniques

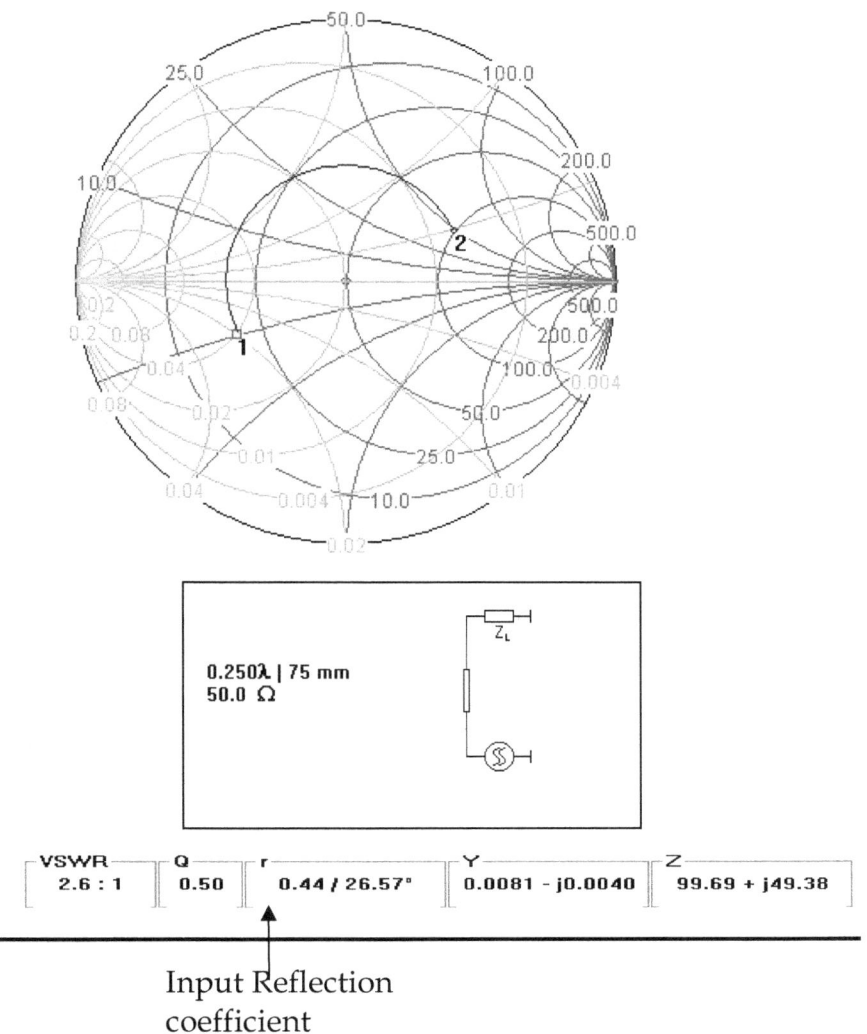

Input Reflection coefficient

A simple rule to remember is that if datapoint 1 is the arbitrary load terminating a transmission line, then moving away from it on the line always results in a ***clockwise rotation on the Smith Chart.***

Lengths of transmission line can be used instead of lumped elements in a circuit for impedance matching or otherwise. Shunt transmission line stubs are an effective way of producing inductive and capacitive effects.

The two kinds of shunt stubs are *open circuited* stubs and *short circuited* stubs. These are shown schematically below.

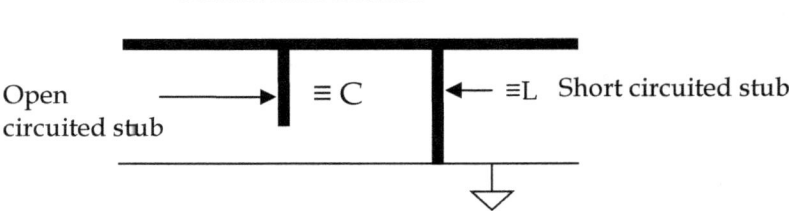

Some interesting facts about terminated and non terminated transmission lines are listed below.

TR1. Very short (with respect to the wavelength of the signal) transmission lines, terminated with a low impedance behave *like inductors.*

TR2. Very short (with respect to the wavelength of the signal) transmission line segments terminated with high impedances or open circuits behave *like capacitors.*

The following expressions can be used to understand the effects.

Practical Impedance Matching Techniques

The input impedance of a transmission line terminated with load ZL is given by: (*Where Zin = R + jX*).

$$Z_{IN} = Z_o[Z_L + jZ_o\tan\theta]/[Z_o + jZ_L\tan\theta] \quad S(18.0)$$

If $Z_L = 0$, then Z_{IN} becomes,

$$Z_{IN} = Z_o[jZ_o\tan\theta]/[Z_o] \quad (S19.0)$$

or:

$$Z_{IN} = jZ_o\tan\theta \quad (S20.0)$$

This is purely reactive. Lets call the impedance of the line Zss for a short circuited segment and its electrical length θss. In this case,

Xss the reactive part of the input impedance = $Z_{ss}(\tan\theta_{ss})$.

The input susceptance is simply the reciprocal of the reactance,

or:

$$B_{ss} = 1/[Z_{ss}(\tan\theta_{ss})] \quad (S21.0)$$

Now if the termination is an open circuit we get:

$$X_{os} = -jZ_{os}/\tan\theta_{os} \quad (S22.0)$$

or, the susceptance is,

$$B_{os} = \tan\theta_{os}/Z_{os} \quad (S23.0)$$

The conclusion is that when the electrical lengths of shorted and open stubs are less than 90 degrees, they behave like shunt inductors and capacitors. The sign of tanθ changes repetitively through every 90 degrees. This implies that both open and shorted stubs can look like inductors or capacitors depending on their electrical lengths.

Transmission line facts: (Repeated for convenience)

Transmission lines that have electrical lengths *less than 90 Degrees* behave *inductively for short circuited loads* and *capacitively for open circuited loads.*

A RF short circuit can be produced *at any point* in a circuit by using a *short circuited transmission line with a half wavelength electrical length.* This short repeats itself every multiple of a half wavelengh.

A transmission line's input impedance is always equal to the termination at adjacent ends of the line if the characteristic impedance of the line is *the same as* the termination.

Transmission lines have *large transformation capabilities.* A short circuit can be transformed to an open circuit by using a *90 Degree* long line. The same is true of an open circuit. An open circuit can be transformed into a short circuit by a *90 Degree* line.

Cascaded transmission lines form *concentric circles* on a normalized Smith Chart if the load impedance is normalized to the characteristic impedance of the transmission line.

Parallel open and short circuited stubs behave *inductively or capacitively* as long as their electrical length is *less than 90 Degrees*.

Parallel stubs *always* move on the constant conductance circles on the Smith Chart.

A correctly selected combination of a *parallel stub and cascade transmission line* can be used to *transform any point* on the Smith Chart to any other point. The topology of this combination is dependent on the relationship of the two points. Sometimes the cascade line can be used first followed by the stub, while at other times the stub is followed by the cascaded line.

The Immitance Chart:

It is possible to use both series elements and shunt elements on the Smith Chart. In order to do so, one needs either, (1) an impedance chart for the series elements and (2) a separate admittance chart for the shunt elements.

What is actually done is that the impedance and the admittance charts are combined. This combinational chart is called an *Immitance Chart*. Generally immitance charts come in two colors to reduce the possibility of error. The impedance chart is drawn in one color while the admittance chart is drawn in a different color. Note that the examples given above are also on the immitance chart. (Except the color is simply black and white)

On immitance charts, impedance movements of ideal lumped elements always occur on the constant reactance or constant resistance circles that intersect the starting impedance. Admittance movement always occurs on constant susceptance or constant conductance circles that intersect the starting point.

Once inside the immitance chart, only *negative resistance* or *negative conductance* can take us *outside* the chart. There may be values that can take us to the circumference of the chart, but only negative values of resistance or conductance can take us beyond that limit.

On the circumference of the chart, the reactance is always equal to the reciprocal of the susceptance. On the main diagonal of the chart the conductance is always equal to the reciprocal of the resistance.

In series circuits, when the circuits have lossless elements such as inductors and capacitors, the immitance chart plots fall on the *circumference where the resistance is zero*. When the circuit has losses, the magnitude of the reflection coefficient is less than unity and the movement occurs on constant resistance or constant conductance circles.

Similarly in the case of parallel circuits, movement is on the constant conductance circles for lossy circuits and on the circumference for lossless circuits.

In order to more fully understand operations using the Smith Chart the following examples are presented. Note that these examples are presented using a Smith Chart program from the web. The idea is to present the technique. There is no need of the software and these manipulations can be done by hand. However, the question is – why? Therefore the approach that we have adopted here is to use software to do the more mundane tasks like normalization and drawing.

Example 1.

Find ZL when the reflection coefficient is found to be 0.5 (from a Vector Network Analyzer for example). The reference impedance is 50 Ohms. The frequency is 500 Mhz. Obviously one can use the equation for the reflection coefficient,

$$\Gamma = [ZL - Z0]/[ZL+Zo]$$

or software to do this. Using the Smith Chart software given provides us with the following figures and solutions.

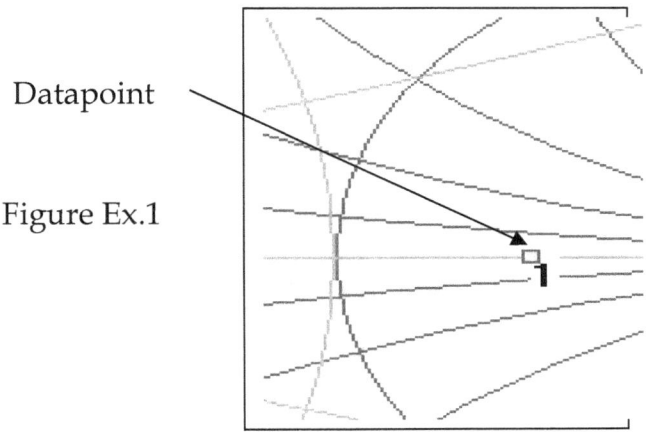

Datapoint

Figure Ex.1

The solution of the problem in example 1.0 was generated by using the Smith Chart software, inserting a datapoint at the value of the reflection coefficient specified and reading the impedance from coordinates generated by the software. Thus when the reflection coefficient is 0.5 with an angle of 0.0 degrees the impedance is 150.0 Ohms as can be seen above.

Further results indicate a VSWR of 3.0 and an admittance of $0.0067 - j0$.

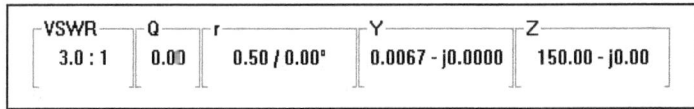

VSWR	Q	r	Y	Z
3.0 : 1	0.00	0.50 / 0.00°	0.0067 - j0.0000	150.00 - j0.00

Example 2:

Find the load impedance when the reflection coefficient is $-0.3 + j0.4$. Here we first convert the reflection coefficient value to polar coordinates as shown below:

The polar coordinates are shown in the results obtained from a *converter script.* Using these coordinates and the Smith Chart program we get following results.

Practical Impedance Matching Techniques

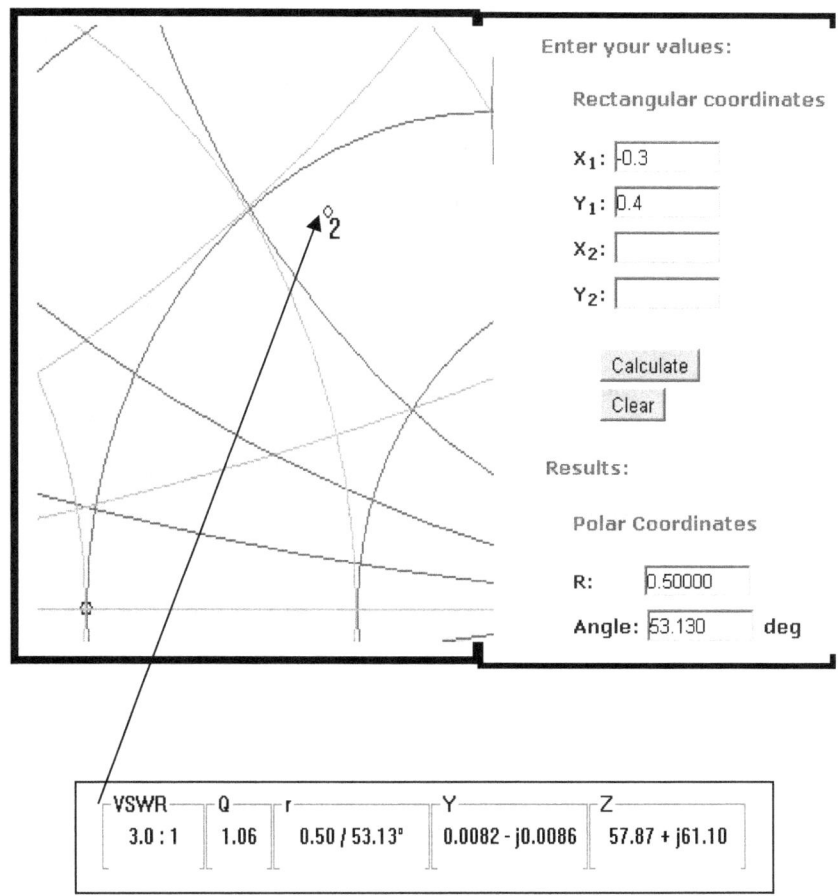

Figure Ex 2.0

Example 3:

Match 100 Ohms to 50 Ohms at 500 Mhz.

A 50 Ohm source needs to be matched to a 100 resistor. The simplest way to do this would be to put 100 Ohms in shunt. However, this would mean unnecessarily dissipating power. Therefore we use "lossless" inductors and capacitors to do the matching.

Point 1 is at 100 Ohms. Point 3 is at the match point i.e. 50 Ohms. In this matching approach, first a capacitor was added, (point 2) then an inductor. The circuit is shown below. The first move was on the admittance chart. The second move was on the impedance characteristic.

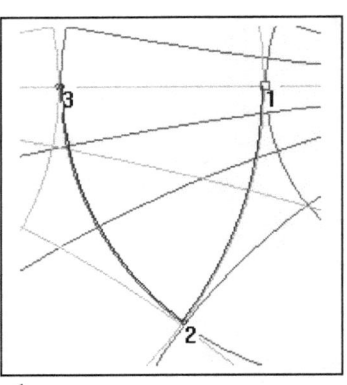

Figure Ex 3.0

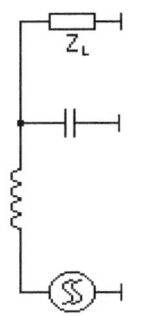

The capacitor is 3.2pF and the inductor is 16 nH.

It should be obvious that another approach will also work. In this case however, the circuit becomes high pass and DC information cannot be passed between the load and source. The circuit is shown below.

Practical Impedance Matching Techniques

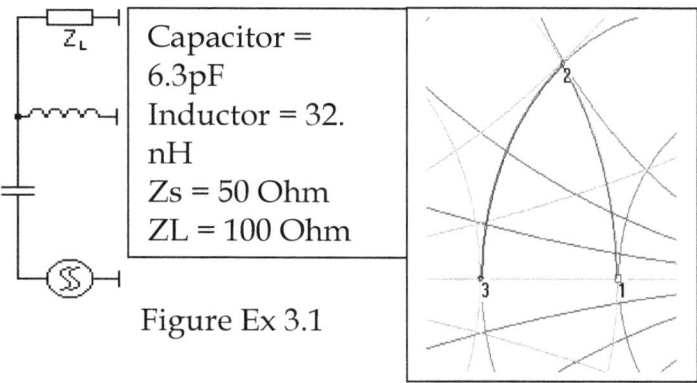

Capacitor = 6.3pF
Inductor = 32. nH
Zs = 50 Ohm
ZL = 100 Ohm

Figure Ex 3.1

Example 4.0:

Match a 24.97+j72.93 impedance to 50 Ohms.

This solution is shown below using the Smith Chart. There are also two moves here. The matching circuit is also shown below.

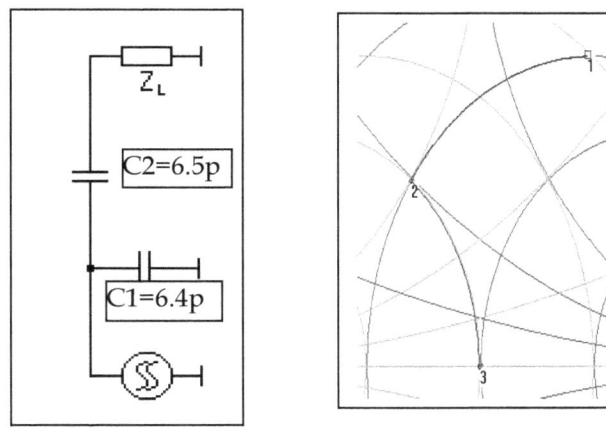

Figure Ex 4.0

These examples serve to show the process of matching using the immitance chart. The source impedance and the load impedance is specified. In the following examples the cascade transmission line is used to show matching techniques.

Example 5:

This example shows how to use a combination of reactance and a cascade transmission line to provide matching. A RF power transistor operating at 1 Ghz has an equivalent input circuit of a resistor of value $Rs = 2.5$ Ohms in series with an inductor of $Ls = 0.88$ nH. This impedance needs to be matched to 50 Ohm. We will find the matching circuitry required to do this job.

The reactance of the 0.88 nH inductor is: 5.5 Ohms. Using these values the following matching manipulation was constructed on the immitance chart.

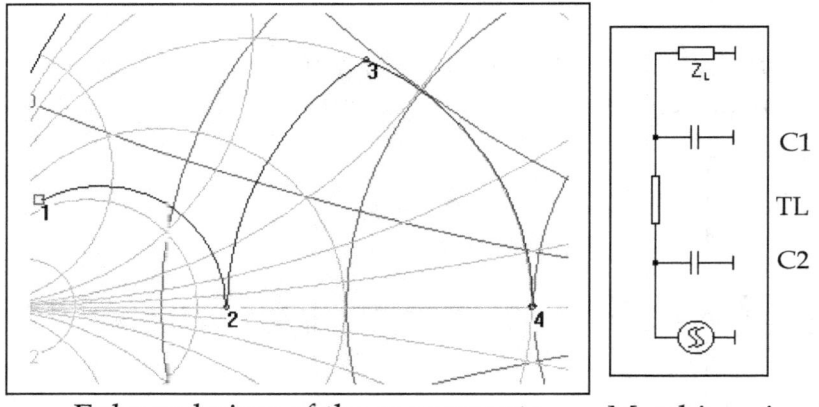

Enlarged view of the movements   Matching circuit

Figure Ex 5.0

C1 = 4.1 pF
C2 = 23.8 pF
TL = 50 Ohm, 0.078λ, 23mm

Example 6:

In this example we explore the use of the Smith Chart to transform an impedance ZL. We will use two lossless transmission lines to do this. The characteristic impedance is 75 Ohm for both lines. The impedance of one port is $30 - j15$ Ohms (i.e. the load).

One solution to this is shown bellow:

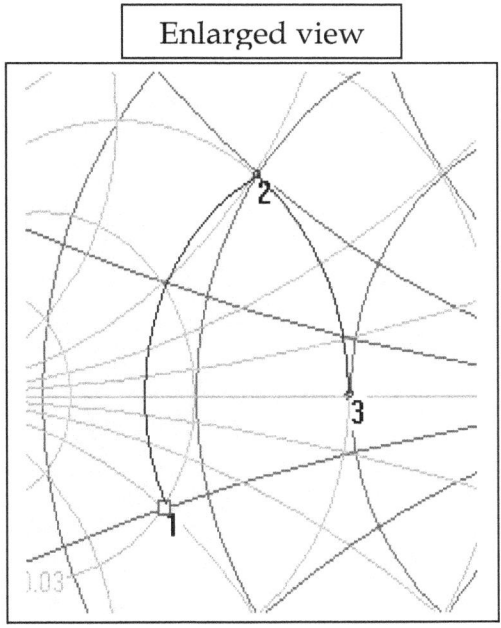

Figure Ex 6.0

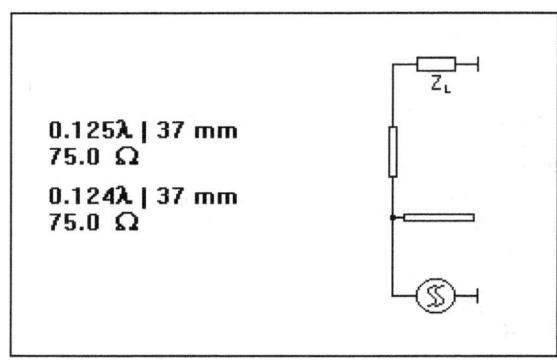

Figure Ex 6.1

As can be seen the transformation was done by using a cascade line and a open-circuited parallel stub with an electrical length of less than 90 Degrees.

It should be obvious that there are other configurations to do this. An example is presented below.

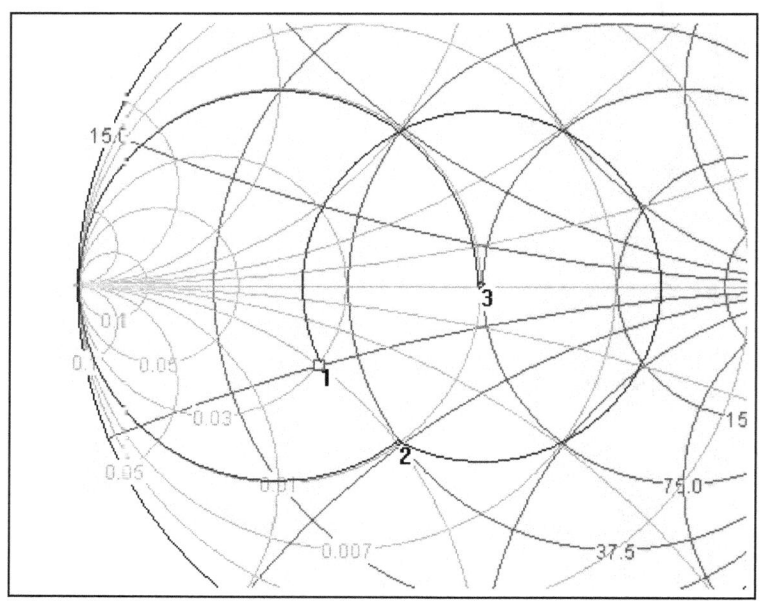

Figure 6.2

The only issue here is the length of the lines! The first configuration has shorter lines.

Single stub matching technique:

Refer to Figure M.8.0 below. It shows the technique used to do impedance matching using a single stub connected in shunt with a main line as shown.

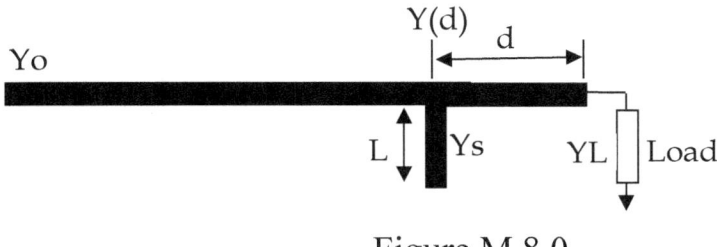

Figure M.8.0

Figure M.8.0 depicts the layout of a single stub matching circuit. The load is YL. The microstrip line has a characteristic admittance of Yo. The load needs to be matched to the characteristic admittance (or the characteristic impedance of the line). A single stub is used in this case that is situated a distance 'd' from the end of the line where the load is connected.

It is well known (as discussed above) the admittance changes as the line is traversed from the load end towards the source. What we want to do is to find a length of line 'd' that will change the load, to where, at the point 'd' it will see the characteristic admittance and a susceptance which can be positive or negative.

In other words at point 'd' we want the admittance to be Yo ± jB. Then we want to connect a stub at 'd' such that its susceptance is opposite in sign to the susceptance generated at 'd'. When this is done the admittance seen at the point 'd ' is Yo and the load has been matched to the line. The following is a description of the method to do this using the Smith Chart.

Lets match a load 25 – j50 using a stub to a 50 Ohm line.

Step 1.0 Put the load admittance on the chart at point DP1.

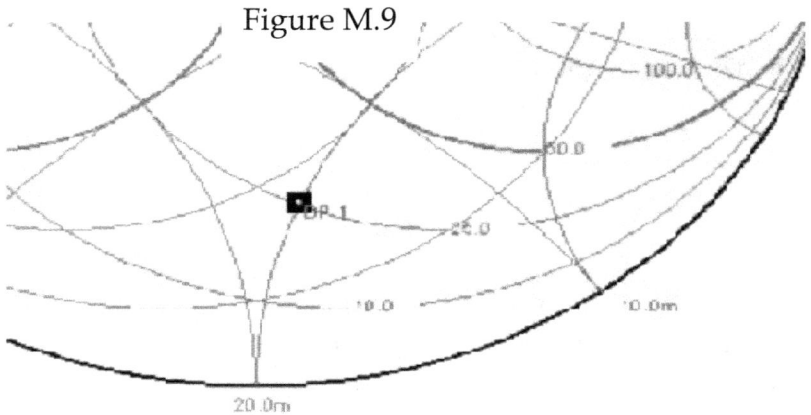

Figure M.9

Note that the chart was set up so as to use admittance and cartesian coordinates. The reason is that shunt stub tuning is being used, so admittance is a more practical choice. So use the immitance chart and set point DP1 as the load as shown.

The details of DP1 are shown below in Figure M.10

Practical Impedance Matching Techniques 147

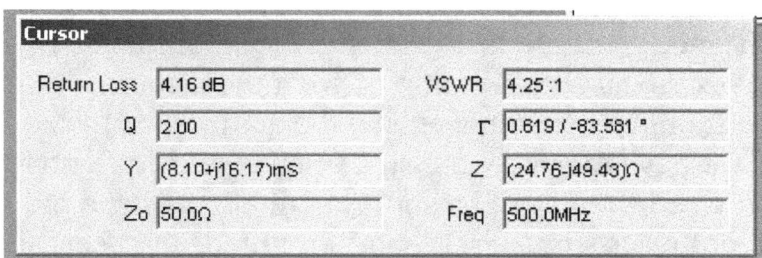

Figure M.10

Step 2.0   Draw a VSWR circle that passes through DP1 as shown below in Figure M.11. (Circles are available under 'tools').

Note that the VSWR circle is the circle that shows the variation of impedance around the chart, Note also that this circle intersects the unitary conductance circle (where the admittance is Yo, i.e the reference admittance of 1/50 Siemens. At both intersection points the real part of the impedance becomes the reference admittance while the imaginary part is the susceptance at that point on the line.

Lets examine this a little further. As we move from the load end of the microstrip the admittance starts changing. At a point where the distance from the load end is 'd' the admittance is Yo $\pm jB$. This gives us our first solution of what the distance 'd' is. Now, in order to have a match we need the admittance to be only Yo. So we have to cancel the susceptance term. We do this by using a stub. This stub can be short circuited at its end or open circuited. Lets assume it is a short circuited stub. Lets start at the short circuited end. At the short circuited end the conductance is infinite. This point is at the left extremity of the center diagnol of the chart. So we will start there. Now because we are dealing with only a susceptance term we will move on the periphery of the chart. The data presented below shows what the numbers are: The susceptance at point A is: 31.47mS. We have to cancel this susceptance by making the stub length and type to be -31.47mS.

Figure M.11

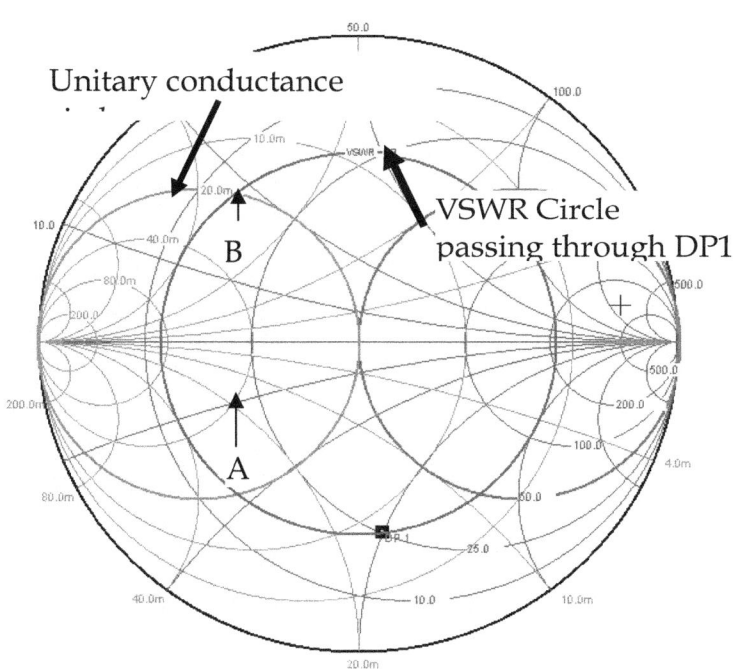

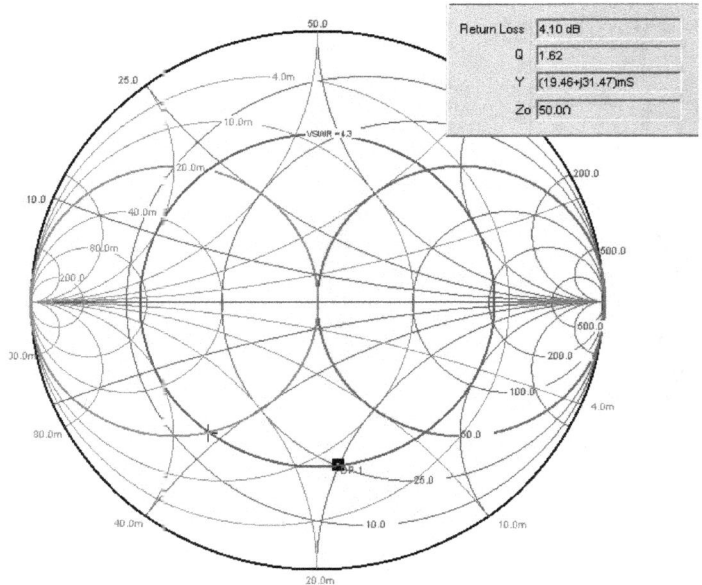

Figure M.12.

A note on the type of stub selected: The analysis and example was based on a shorted stub. You can use an open stub. What kind of stub to use depends
on preference and ease of use and tuning and modifying. If you use an open circuit stub then you have to start from conductance of zero (at the right hand side of the chart). Please refer to chart properties in the section on Smith Chart properties.

Practical Impedance Matching Techniques

Angle is 123.94

Stub length is: 180 − 123.94 Deg = 56.06 Deg. Move clockwise around the periphery towards the generator (or source) Verify this using the equation for a shorted stub.

At this point the susceptance is -37.57mS

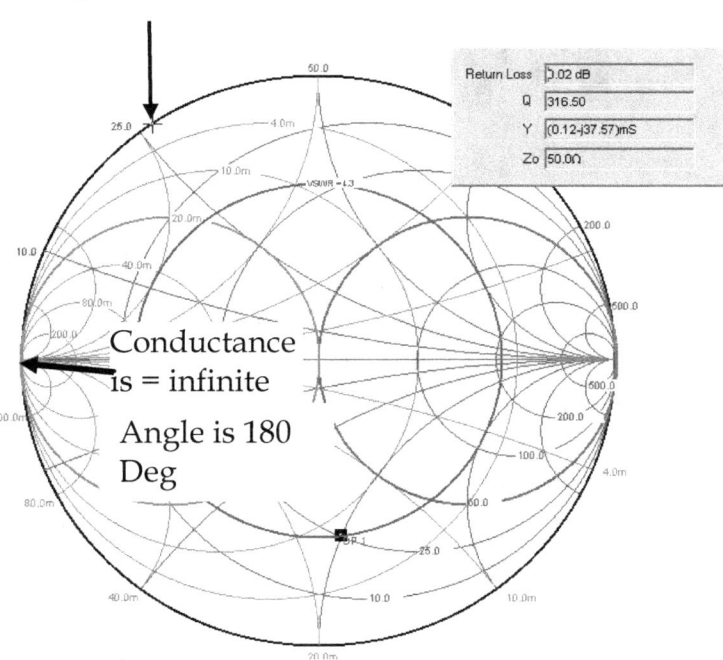

Conductance is = infinite

Angle is 180 Deg

Figure M.13

Double stub matching technique: Single stub tuning has advantages and disadvantages. It is simple to understand and use. However, if the load impedance changes then we have to adjust 'd' every time. This can be quite tedious. In double stub tuning (Figure M.14) the distance between stubs and 'd' can be made fixed and only the stub lengths changed to tune a wide band of loads. However, even double stub tuning has the disadvantage of *not* being able to match all impedances and we must resort to triple stub tuning. In this section of the book we examine double stub tuning to understand what it is and how it is done. The advantages and disadvantages will be made clear as we work through this technique of impedance matching.

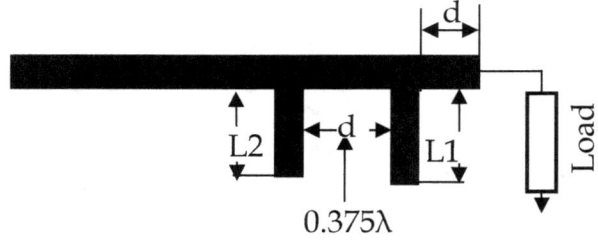

Figure M.14

Double stub matching: Double stub impedance matching is more involved than single stub matching. First of all we need to discuss the rationale for using double stub matching. In order to understand all this, consider the points below:

1.0 A single stub match involves the use of a stub of fixed length placed at a fixed position on the transmission line from the load for a specific load.

2.0 If the value of the load changes then the length of the stub and the position at which it is placed must also change.

Practical Impedance Matching Techniques

3.0 A better technique would be where we could fix the positions of the stubs in relation to each other and only change the lengths of the stubs to match varying loads. This is done using the techniques of double and triple stub matching.

4.0 Both analytic and graphical techniques are available to do double stub matching. The graphical method is usually the Smith Chart method and that is what we will focus on in this discussion.

5.0 In order to further understand the technique of double stub matching we will briefly digress to discuss some aspects of the Smith Chart that will clarify the technique for readers.

6.0 There are some loads that cannot be matched using double stub matching by simply altering the length of the stubs. However if we are willing and able, to move the stubs together a distance away (or towards the load) then we can accommodate the load. This will be discussed and illustrated using the Smith Chart method. These loads form an area of the Smith Chart collectively known as the *forbidden zone* for double stub matching. For example, if the stub spacing is $\lambda/8$, $3\lambda/8$ or $5\lambda/8$, then the forbidden zone is the entire area of the Smith Chart encircled by the $g = 2$ circle. If it is $\lambda/4$ then the forbidden zone is the area surrounded by the $g = 1$ circle. For more information see Ref SC.1.

7.0 Triple stub matching can be used to overcome the limitations of double stub matching if needed. Similar techniques to that of double stub matching are used for triple stub matching. However triple stub matching is not commonly used in microstrip circuits.

Background for the use of double stub matching:

8.0 *The microstrip (or transmission line) as a transformer.*
A length of microstrip with a characteristic impedance Zo acts as a transformer. To visualize this, we can look at the following Smith Chart plot that shows this. In addition, note that the Smith Chart circumference is a measure of the distance moved on a transmission line. For one complete revolution we travel half a wavelength on the line. From the chart its easy to see that a movement from point 1 to point 2 changes the imaginary component (reactance or susceptance) of the line. This is a transformation. So as we move on a transmission line, the line acts as a transformer. This fact is used very successfully in the design of the popular quarter wave transformer.

> *Note that one complete rotation around the Smith Chart is one half wavelength on the line. This is also represented by 360 degrees around the Smith Chart.*

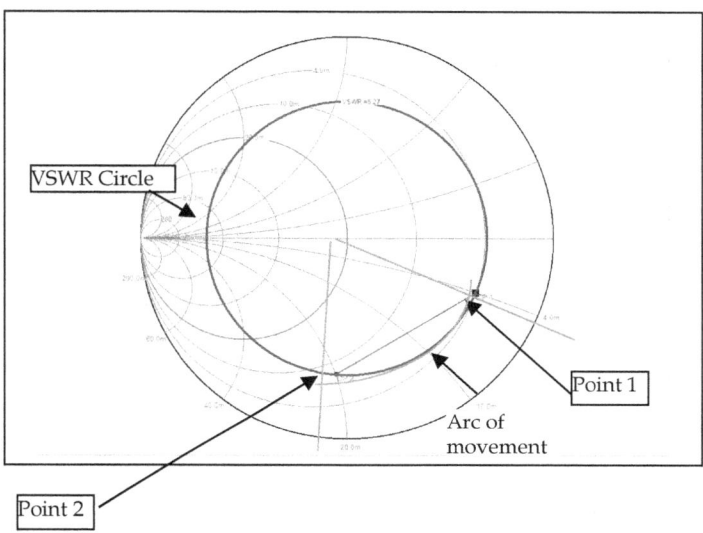

9.0 *The spacing circle.* In double stub matching the two stubs are spaced a predetermined distance away from each other. These distances are typically $\lambda/8$, $\lambda/4$, $3\lambda/8$, $5\lambda/8$ etc. Knowing what we know about a length of line acting as a transformer, we know that the length of line between the two stubs acts as a transformer. *The action of this transformer is to convert the admittance at the position of stub 2 to a different admittance at the* **position of stub 1.** So we *start* from the position of *stub 2.*

In order that stub 2 can be finally used to match the line admittance, the real part of the admittance at the position of stub 2, on the line, has to be 1.0 (normalized value). Its susceptance is then jB. jB is the susceptance that is cancelled using stub 2 to ultimately get the matching to the line admittance. The admittance at the position of stub 2, (without the stub) lies on the <u>constant conductance, g = 1 circle</u>. The admittances on the g = 1 circle are all the <u>possible admittances</u> at the stub 2 position for match to take place.

To reiterate, as a result of these deliberations, that some point of the VSWR circle formed by the position of stub 2, ***must*** intersect the g=1 or the unity conductance circle on the Smith Chart.

We also conclude, that the admittance at the position of the first stub, must lie on a circle of *equal radius* but having its center rotated (moved to, or displaced) by the spacing between the stubs *towards the load.* Lets call this spacing 'x'.

This circle is called the *spacing circle* and its construction is explained further below. The concept of the spacing circle is important in understanding the graphical method of double stub matching using the Smith Chart. [Note that *one complete rotation around the Smith Chart is <u>360 Degrees</u> for $\lambda/2$.* This fact allows us to graphically construct movements on the line on a Smith Chart. For example a $\lambda/4$ length of line is a movement of 180 Degrees on the Smith Chart.]

Consider the Figure shown below where we have assumed a stub spacing of λ/4.

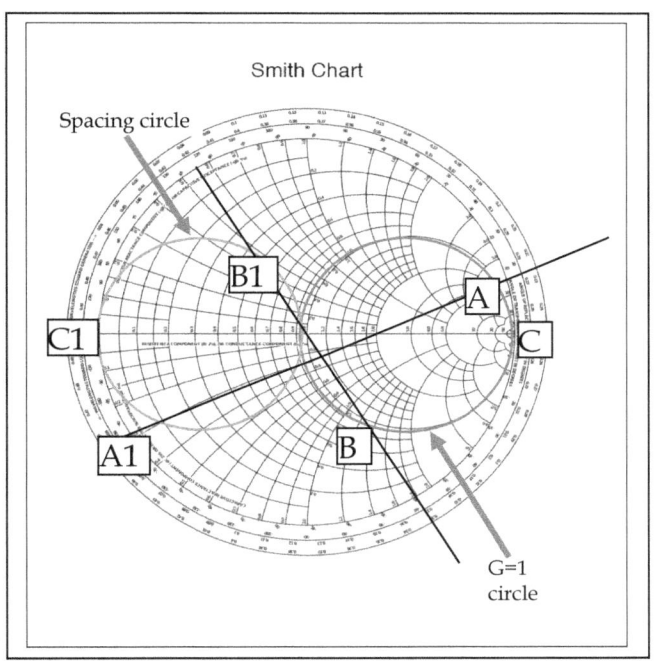

Note that a spacing of λ/4 implies a rotation or displacement by 180 Degrees.

In this figure points A, B and C have been "rotated" or moved by 180 degrees to new positions A1, B1, C1. In the same way *all points* on the first circle on the right can be moved to the circle on left thereby accomplishing *the rotation* of the circle by λ/4. This also shows how the impedances (or admittances as the case may be) change with distance of movement on the transmission line between the two stubs. A *transformative property* of the transmission line.

So it can be seen, that once we know the distance between stubs, the admittances on the line can be transformed by a movement or *spacing, towards the load* as shown. The admittances lie on a circle of the same radius but displaced *towards the load*. *Any* spacing between stubs can be handled in this way and spacing circles formed. The center of the spacing circle has been rotated by the length of line between stubs. The radius is still the same only the center has moved to a new position dictated by the distance between the stubs.

Similar transformations can be made for other values of x. In every case the locus of points on the unit conductance circle maps into a new *spacing circle* of the same radius, whose center lies x wavelengths *towards the load* (counter clockwise) from the center of the unit conductance circle.

Any intersections of the original VSWR circle with the g=1 circle also are transformed by this procedure and now lie on the spacing circle. In this way we see how the length of line between the two stubs affects the admittances.

To reiterate: To develop a spacing circle simply move the reference circle that lies at a conductance of g=1 and move the points on it by the angular distance between the two stubs to form the *spacing circle*.

The importance of this technique in the double stub matching process cannot be overemphasized. It forms the crux of the technique and should be understood to an *intuitive extent*, by the engineer who is interested in graphical methods of stub tuning using the Smith Chart.

Practical Impedance Matching Techniques

A further important concept in double stub matching is the *forbidden region*. This is a region on the Smith Chart which consists of impedances that cannot be matched. In his original work Smith describes these regions based on some discrete stub spacings.

The analytical treatment of forbidden regions is somewhat involved but to state the results let us define *gload* as the conductance of the load. Then the forbidden region may be defined by the following constraint:

$$0 \leq \text{gload} \leq 1/\sin^2\beta x$$

where,

x = spacing between stubs
β = wavenumber = $2\pi/\lambda$

i.e The forbidden region is surrounded by a constant conductance circle whose value depends on the electrical stub separation, d/λ. For example, lets assume that the separation between the stubs,

$$x = 3\lambda/4$$

so,

$$\beta x = 3\pi/4$$

or,

$$1/\sin^2\beta x = 2.0$$

Its clear from this, that the forbidden region is enclosed by the gload = 2 circle on the Smith Chart.

These identities can be used to define the forbidden regions on the Smith Chart for various loads.

Modification of double stub matching to override the forbidden region restriction:

If we find that the load admittance lies in a forbidden region we can still use double stub matching if we insert a length of line between the load and the first stub of length = *lenz*.

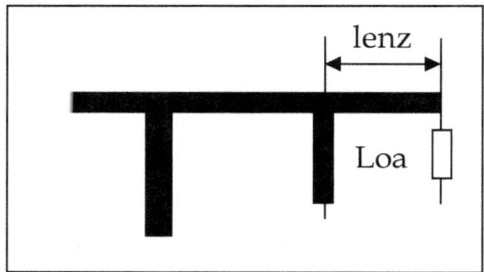

The effect of the added length is to change the load admittance so that the transformed value of *gLoad* is *transformed* to a level that meets the requirements of double stub matching. See the figures below.

Practical Impedance Matching Techniques

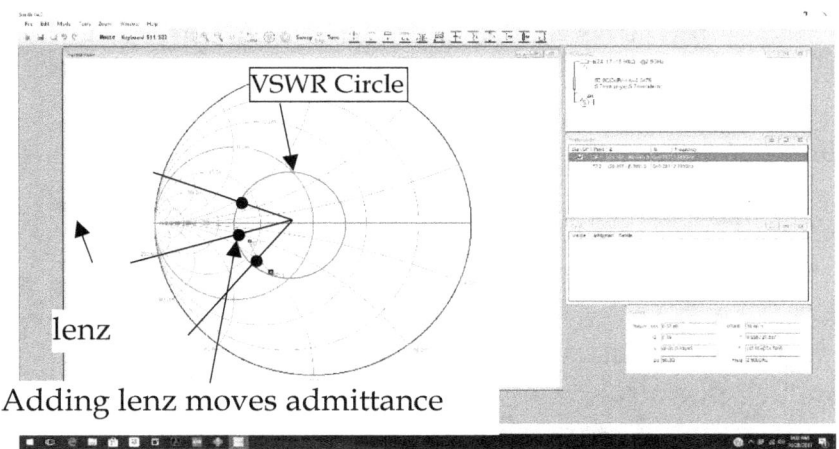

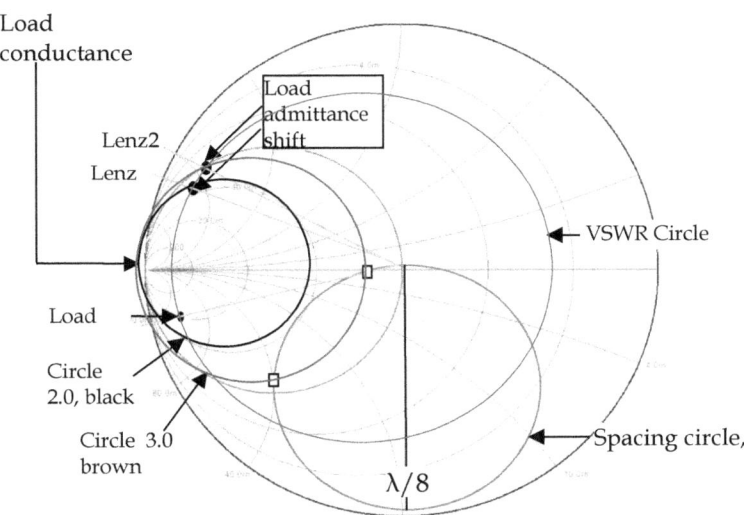

Generate the spacing circle with a shift of λ/8

Points of intersection of the conductance circle (circle 3) with the spacing circle when lenz=lenz2 in yellow (small boxes)

Notes on the two figures above.

Practical Impedance Matching Techniques

1.0 The first figure shows that as a length of microstrip (or transmission line) is added, the admittance moves to a new value on the VSWR circle. A transformative function.

2.0 The second figure is much more involved. We will break it down as shown below:

3.0 An admittance is shown on a Smith Chart along with its conductance circle shown in green. For black and white diagram the arrow should distinguish it from other circles.

4.0 It is assumed that the two stubs are spaced by $\lambda/8$. As a result the spacing circle is rotated by this amount and is shown in red (along with its caption with an arrow).

5.0 Lengths of line, lenz1 and lenz2 are added sequentially. These line lengths move the admittance values on the VSWR circle shown in brown (or arrow).

6.0 It can be seen quite plainly that the non transformed admittance lies on a conductance circle which *does not* intersect the rotated spacing circle at any point. As a result the two stubs cannot be used for matching this admittance.

7.0 As the admittance is moved by the addition of lenz1 and lenz2 it is found that the conductance circle for lenz2 intersects the spacing circle at two points. Therefore matching can be done using the double stubs.

8.0 If the first stub _has_ to be right at the position of the load, then the distance _between_ the two stubs has to be changed such that the conductance circle of the load and the spacing circle _does_ intersect.

Applying these concepts to double stub matching:

Lets apply these concepts to double stub matching. Refer to the figure below which shows the two circles. The $g = 1$ circle and the $\lambda/4$ spacing circle to start with.

The blue circle represents stub 2 position without the stub present. The red circle represents the spacing circle for a quarter wavelength spacing of the stubs. It also represents the admittances that are the result of the transformation effected by the length of line say, $x = \lambda/4$.

In the next figure we place the load admittance on the chart as shown at point P1.

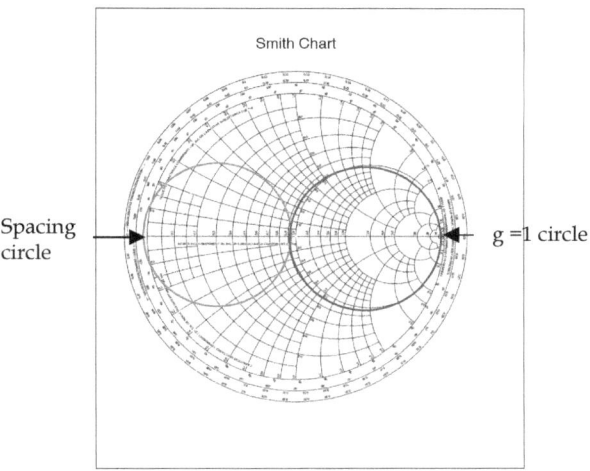

The g = 1 circle (blue) and the spacing circle for quarter wave length of spacing between the stubs is shown.

Its interesting to note that the spacing circle and the load conductance circle represent what is happening at the *load end* of the circuit. There are two points of intersection of the load conductance circle with the spacing circle, P2 and P3. We can use either to perform the matching.

In this case let us use P2. To understand what is happening here note that the spacing circle and the load conductance circle represent the action *at the load end* of the circle as stated above. What we want to do is to move the load admittance to fall on the point P2. This point, when looked at *from the stub2 end* (by virtue of the transformation property of the line) will cause it to fall on the g=1 circle as needed by the matching requirement. This act represents the reverse of the spacing circle move. Now we are moving from the spacing circle to the g=1 circle. This is of course very valid; it's a reciprocal relationship as noted in previous discussions above. The next figure and its associated discussion shows how this is done. Θ represents the angular movement of the point P1 to point P2.

This can be accomplished by adding a stub with the correct value at the load. The correct value is the *difference between the susceptance at point P1 and P2 or P3.*

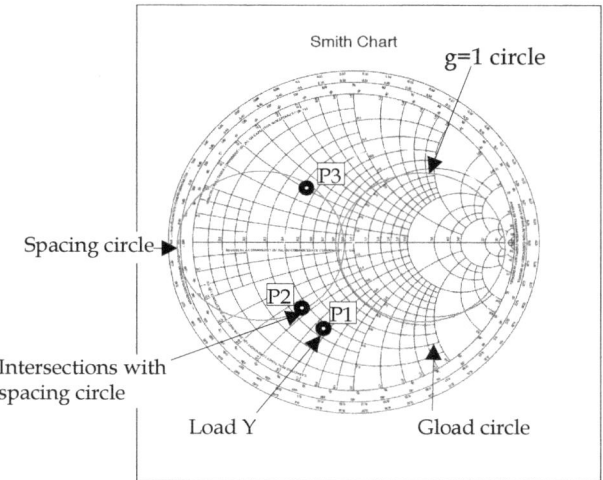

Once the stub is added to the load side of the line it moves P1 to P2 (or P3) and then the transformation to the stub 2 side shows that the conductance is now unity with a susceptance component. The susceptance component is cancelled using stub 2 with an equal and opposite susceptance.

At this point matching is complete.

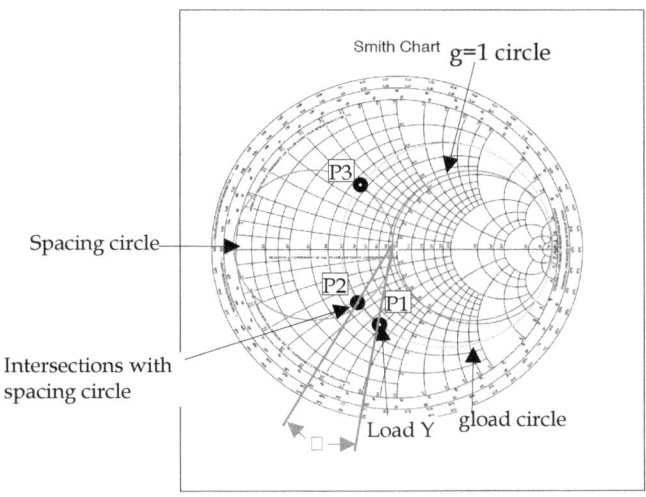

*Please note that this book does not address the triple stub matching technique at this time. Triple stub matching can be very useful in a similar way to overcome forbidden regions and thereby match the loads. However, it is not commonly used in microstrip type design.

Example of double stub matching.*

- A good example of double stub matching with simulation results and schematics is provided in a companion book by the author: " Impedance matching using double stubs" available on Amazon at :

- https://www.amazon.com/s/ref=nb_sb_noss?url=search-alias%3Dstripbooks&field-keywords=impedance+matching+using+double+stubs

Part VI
Appendices

dBm: Power ratio in dBs with the reference power level of 1 mW.

$$Po(dBm) = 10Log(Pin/1e-3)$$

Po = Power in dBm
Pin = Power in Watts

Voltage can be converted to dBm. For these conversions the reference resistance is required, as shown below.

$$P(dBm) = 10log(v^2 /[R*1e-3])$$

R is the reference resistance or load resistance, typically 50 Ohms in RF systems.

v is the rms voltage.

Vrms = 0.3535* V(peak to peak)

There are many dBm to Vrms and vice – versa converters on the World Wide Web. A converter is also available in the calculators accompanying this book.

Details of the Smith Chart Software used in this eCADbook.

V 1.91
This program has been developed by Prof. Fritz Dellsperger, Juerg Tschirren and Roger Wetzel
© 1995 - 2000 by Berne Institute of Engineering and Architecture

─ Licence ─
No valid licence. This copy of 'smith.exe' runs as a DEMOVERSION.

It was downloaded from the web as freeware. Permission for its use was granted by the authors.

A number of calculators and converters were used to verify the theory presented in this book. These were all available on the World Wide Web as well as in the calculators accompanying this book.

- Convert rectangular to polar coordinates
- L and C reactance calculators
- Calculators for complex numbers

References

1.0 Circuit design using personal computers. Thomas R. Cuthbert Jr., Wiley – Interscience Publications, John Wiley & Sons.
2.0 Fields and Waves in Communication Electronics, Simon Ramo, John R. Whinnery and Theodore Van Duzer. John Wiley & Sons, Second Edition.
3.0 Practical RF Circuit Design for Modern Wireless Systems. Volume I, Les Besser and Rowan Gilmore, Artech House.
4.0 Foundations of Interconnect and microtrip design. T.C. Edwards, M.B. Steer, third edition, J. Wiley & Sons, Ltd.

5.0 A study of microstrip design in silicon technology using closed form analytical expressions. Ain Rehman, Signal Processing Group Inc.
6.0 Electronic applications of the Smith Chart, Phillip H. Smith, Noble publishing.

Javascripts and other useful programs for calculating many of the quantities described in the book are available from the publisher. For further information please contact the publisher at : spg@signalpro.biz

Notes:

www.ingramcontent.com/pod-product-compliance
Lightning Source LLC
Chambersburg PA
CBHW052256220526
45471CB00001B/365